DARK DAYS AT NOON

Edward Struzik

DARK DAYS AT NOON

THE FUTURE OF FIRE

McGill-Queen's University Press
MONTREAL & KINGSTON • LONDON • CHICAGO

ISBN 978-0-2280-1209-2 (cloth)
ISBN 978-0-2280-1348-8 (ePDF)

Legal deposit third quarter 2022
Bibliothèque nationale du Québec

Printed in Canada on acid-free paper

This book has been published with the help of a grant from the Canada Wildland Fire Network.

Funded by the Government of Canada | Financé par le gouvernement du Canada | Canada

Canada Council for the Arts | Conseil des arts du Canada

We acknowledge the support of the Canada Council for the Arts.
Nous remercions le Conseil des arts du Canada de son soutien.

LIBRARY AND ARCHIVES CANADA CATALOGUING IN PUBLICATION

Title: Darks days at noon : the future of fire / Edward Struzik.
Names: Struzik, Edward, 1954– author.
Description: Includes bibliographical references and index.
Identifiers: Canadiana (print) 2022019484X | Canadiana (ebook) 20220194912 | ISBN 9780228012092 (cloth) | ISBN 9780228013488 (ePDF)
Subjects: LCSH: Wildfires—Canada—History. | LCSH: Wildfires—Canada—Prevention and control—History. | LCSH: Wildfires—United States—History. | LCSH: Wildfires—United States—Prevention and control—History.
Classification: LCC SD421.34.C3 S77 2022 | DDC 363.37/9—dc23

CONTENTS

ACKNOWLEDGMENTS

I have experienced a lot of fire in my time, most notably on canoe trips through the boreal forest of the sub-Arctic where there were no firefighters to extinguish them. Fortunately, most were too far away to do me harm. But one that seemed to be coming at me as I was hiking up the Lockhart River in the Northwest Territories was frightening because of the noxious smoke and because I had no idea which direction to run to escape it. The 2004 fires in the Yukon and Alaska were also memorable for the fact that they were so big and all-encompassing that they put an end to plans for me and my family to drive down the Alaska Highway and canoe the Wind River that summer.

My first meaningful insights into the science and history of fire had just arrived the year before when Parks Canada's Michel Boivin, Mark Heathcott, Rick Kubian, Rob Walker, Steve Otway, and Dave Smith gave me the chance to see how intense, fast-moving fires like those that swept through the Rocky Mountain parks that summer behaved and threatened nearby towns. The fires in Kootenay were caused by lightning. The one in Jasper was a prescribed burn that got out of control. It was still a time when government employees could share their thoughts without having to go up the chain of command where responses to questions would be vetted and answers rehearsed in advance so there would be no controversy.

The year 2003 was what Kubian described as a "career fire season." He and his colleagues had never experienced anything like it before. A long hot summer turned the valleys into furnaces, setting the stage for duff and fire-resistant hardwoods to burn even at night, when fires normally go to sleep and smoulder. Then the winds came in, forcing officials to engage in the potentially career-ending strategies that I describe in this book. In both cases, they were damned if they didn't take the risk and damned if the strategies they proposed failed.

I didn't fully appreciate what Rob Walker meant when he predicted that the 2003 fire season was a "harbinger" of what was to come in a rapidly warming world. But later that year British Columbia and California burned big, forcing the kind of mass evacuations that had been rare in the previous half-century.

The record-breaking 2004 wildfire season in the Yukon and Alaska was followed by a series of truly remarkable fires such as the Anaktuvuk fire that incinerated so much tundra on the North Slope of Alaska in 2007.

In 2008, Brian Stocks of the Canadian Forest Service invited me to come to Cold Lake, Alberta, where he and 120 scientists from around the world were participating in a $24 million NASDA-led project tracking smoke and the eruption of pyroCbs. Stocks has spent an inordinate time helping me understand the science of fire, as has Mike Flannigan, a meteorologist and fire scientist who has encouraged me to no end in this book-writing effort. Cliff White, the architect of Parks Canada's wildfire management plan, has also shared many insights and physically walked me through a future fire scenario in Banff.

There are many other fire scientists to whom I am grateful: Scott Rupp, deputy director at International Arctic Research Center, University of Alaska Fairbanks; Andrew Larson from the University of Montana; Jill Johnstone from the University of Saskatchewan; Jen Beverly from the University of Alberta; Marty Alexander, Ellen Whitman, Mike Norton, Roger Brett, and Brian Simpson from the Canadian Forest Service; and Uldis Silins, who explained the research he has been doing in the aftermath of the 2003 Lost Creek fire in Crowsnest Pass and the 2017 Kenow fire in Waterton Lakes National Park. I also had a memorable day with Chad Hanson when we hiked through the Sierra Nevadas in California where the Rim Fire burned big in 2013.

Were it not for Fred Wurster, a US Fish and Wildlife Service scientist working in the Great Dismal Swamp National Wildlife Refuge, I would never have known that some firefighters wear hip waders and that fire is a regular visitor to degraded wetlands. It was a pleasure keeping up with Mike Fromm when he was affiliated with the United States Naval Research Laboratory unravelling the mysteries of pyroCbs.

Another side of the fire story involved botanists and biologists such as Gordon Stenhouse, Elly Knight, Scott Nielsen, John Spence, Federico Riva, and many others who shared their insights into fire's impact on plants and animals. My five expeditions with Gordon capturing grizzly bears were easily the most memorable of the field trips I did.

I would also like to thank ecohydrologist Mike Waddington for guiding me through the fire that burned around the wetlands in Georgian Bay in 2018. And to the folks at Parks Canada who shared photos of fires that burned in Canada's national parks, thank you as well.

Thanks to McGill-Queen's University Press for giving me the opportunity to write this book and to editor Khadija Coxon for guiding the manuscript through the peer review process and clearing the way for Susan Glickman to edit the book. The anonymous peer reviewers invested a lot of time, offering suggestions, making corrections, and encouraging me with very kind words. I figured that Kathleen Fraser, the managing editor at McGill-Queen's, must have thought that the book had merit when she brought Susan Glickman on board. Susan is a very fine novelist, poet, and essayist. It was humbling but gratifying to work with someone with so much talent.

To my wife Julia Parker, who considered my thoughts and read through the manuscript, making invaluable suggestions in both cases, I am also grateful, as I have been in the past when she helped me with other projects.

Finally, another shout-out to Mike Flannigan and the University of Alberta for making it possible to have so many photos published in the book and to Kari Greer, whose fantastic photographs for the US Forest Service grace the cover as well as a number of inside pages.

DARK DAYS AT NOON

INTRODUCTION

During the first hundred years of Canada's history, no wildfire was noted in the books written by Donald Creighton, Harold Innis, W.L. Morton, Frank Underhill, and Arthur R.M. Lower: men who dominated national storytelling until 1967 and for some time after. Creighton's 1972 book, *Canada's First Century*,[1] considered at the time to be a milestone in Canadian literature and scholarship, is a book that I and many other university students were assigned to read in both undergraduate and graduate history classes.

The book is about the French and the English, the so-called "founding people" of Canada. It is about tariffs, taxes, and treaties, the Wheat Board, two world wars, and the British North America Act. Unnamed militant farmers and riotous labourers who tried to upend the natural order of the British institutions that Creighton treasured are acknowledged, but not by name, nor for the inconvenient truths they espoused about Canada's political leaders and the business establishment they were beholden to. There are many powerful men like Macdonald, Mercier, Meighen, and even Mussolini, but no reference to women such as McClung, McKinney, and Murphy who, along with Irene Parlby and Henrietta Edwards, were members of "The Famous Five," litigants in the "Persons Case" of 1927 that was a landmark legal decision for women's rights in Canada.

India, the country, is mentioned six times in Creighton's book. "Indians," the word used to describe Indigenous people back then, are referred to only once as "tribes" that peacefully surrendered their original title to land in exchange for living on reserves. No mention is made of them starving due to the government's wilful neglect, the residential schools their children were forced into, or the sterilizations that tarnished the reputation of the "Famous Five" because of their association with the eugenics movement, which advocated selectively mating people with specific desirable hereditary traits, thereby breeding out disease, disabilities, and in some cases, skin colour.[2] The extirpation of millions of the bison that drove both settler and Indigenous economies well into the first decade of Canada's first century is ignored. In Creighton's book the Inuit do not exist, and the Métis are quickly dispatched in a few pages with the hanging of the rebellious Louis Riel.

Some criticized Creighton's book for its lament for a nation that had turned its back on the British Empire in favour of a continental partnership with the United States. Others disavowed it for its contempt for Quebec nationalists. But as Donald Wright, Creighton's biographer, pointed out, it didn't matter; the book was a hit with the Canadian public and the academic community, initially outselling *Everything You Always Wanted to Know about Sex,* and, for a time, the Bible. This was the Canada that most Canadians knew and understood.[3]

One might have expected Arthur R.M. Lower, renowned for many articles and books on the spoliation of Canada's forests, to describe the role that wildfire played because it often had an enormous impact on timber harvesting, pulp mills, gold mining, and agriculture. Like Creighton, Lower was fiercely nationalistic, so much so that he once told a New York City audience that "It's mistake to think that Canadians don't like Americans. The fact is Canadians hate Americans."[4] During his university study years, Lower spent several summers as a fire ranger in northern Ontario, putting up posters, advising people to put out their fires, and fighting small blazes that ignited when that advice was ignored. He was working in the bush during the summer of 1911 when an enormous fire, not far from where he was camped, torched more than 200,000 hectares (more than three times the size of the city of Toronto) around the Porcupine Lake area near Timmins in northern Ontario, killing at least seventy-three people and possibly as many as 200. He was teaching at a university in Manitoba in 1929 when a forest fire levelled the town of Cranberry Portage 700 kilometres to the north.

He was around as well when fire killed 223 people in Matheson, Ontario, in 1916, at least 13 in Saskatchewan and Alberta in 1919, and 43 in Haileybury-Temiskaming, Ontario, in 1922. The damage done by the Haileybury Fire was so catastrophic, the city of Toronto offered to send seventy streetcars to help house the thousands who were left homeless.

And yet, in spite of all his experience, insight, and knowledge, fire gets only two brief mentions in Lower's most important book, *The Northern American Assault on the Canadian Forest*, which was favourably reviewed by Creighton in 1939.[5] Nothing is said of railroad companies being responsible for many of the

1.1 The Manitoba town of Cranberry Portage burned in 1929. Photo by Francis Roy Brown.

1,500 fires burning annually in Ontario in the first decades of the last century.[6] His passing reference to the 1825 Miramichi Fire, one of the biggest and most deadly in North American history, says nothing about the fact that at least 160 and likely many more people died, or that timber exports plunged by two-thirds over the next decade.[7] The only mention of Miramichi in the main text is a reference to eighteen American lumbermen – "trespassers," as he describes them – who were compensated for their losses by a Canadian relief fund.[8]

Political and economic historians such as Lower and Creighton took a top-down approach that never got to the bottom where plants and animals, forests and fens, drought and floods, wildfires and climate, and other environmental forces have shaped the cultural and economic development of Canada and its evolving relationship with the United States more than they did its devolving relationship with Great Britain. Ancestry, and British institutions such as the monarchy and a parliamentary democracy, link Canada with the United Kingdom. But a big ocean separates them. Nature, and a shared experience of settling into and exploiting the wilderness of the New World, has compelled Canada, like it or not, to partner with the United States, even as "God Save the Queen" echoed daily through the hallways of Canada's schools during its first century and many years beyond. Wildfire was one of those natural, and sometimes not-so-natural, events that shaped that relationship as well as the attitudes, beliefs, and public policy initiatives that came out of them. There was little in the British experience that could teach Canada, or the United States for that matter, much about forest fires in North America because consequential wildfires did not occur over there with frequency, and those that occurred throughout other parts of the British Empire represented a much different challenge when it came to fire management. The London-based Empire Forestry Association, which had as many as 260 members in Canada in the 1920s, was good in advising how to grow and cut down trees. But it and other Commonwealth associations that came before it were clueless when it came to dealing with fire.

Historians were not the only ones who ignored the wildfires that engulfed so much of Canada and the United States. In spite of being one of the four elements of matter, along with earth, water, and air, fire found no place in the science disciplines nor in popular history. Best-selling author Pierre Berton (1920–2004) wrote two wildly popular trade books – *The National Dream* and *The Last Spike* – on the building of the allegedly nation-building transcontinental railroad. Like Creighton, Berton makes no mention of how so much of the country's forests, prairies, and towns burned because of sparking train wheels, embers spewing from the chimneys of wood- and coal- and peat-burning locomotives, and hot ash that was routinely shovelled out and left along the side of the tracks in the middle of sun-baked forests. Entire towns were burned down to the ground in this way.

In his chart-topping book *Caesars of the Wilderness,* Peter C. Newman (1918–1991) wrote colourfully of the fur trade in which hard-drinking voyageurs murdered and ambushed, stole and kidnapped, maimed competitors, and destroyed property. But in the logistics of fur trading – canoeing and portaging along hot, buggy, mossy, grassy, and forested supply lines that were more than 3,000 kilometres long – not one of the more than 200 significant wildfires recorded in the southeast Canadian prairies between 1796 and 1870 gets in the way.[9]

Nor are there Indigenous people lighting prairie grass to herd bison away from the fur traders so that they would starve them out and make them pay more for meat and hides. (The thought of Indigenous people outwitting French and English fur traders never entered the discussion, even though some of the traders reluctantly admitted to it in their journals.) The only mention of fire in Newman's book is by John Henry Lefroy, a Royal Artillery surveyor whose description of a distant forest fire at Methye Portage in northwestern Saskatchewan in 1844 was so quaintly written in the manner of the picturesque that its description borders on the improbable, as did the sketch that artist/explorer George Back drew of the portage in 1825. "A portion of the wood in the distance was burning, and there was an uncommon felicity in the manner in which the columns of smoke rose up against a dark mass of Pines, which cross the valley behind them."[10]

I.2 *Moose Attracted by a Forest Fire at Night*, watercolour by Denis Gale painted circa 1860.

Novelists, poets, and artists tend to be ahead of the curve, revealing rather than ignoring the realities of the world. But up until, and well beyond the time the Group of Seven painters began venturing into clear-cuts and the burned-out forests of northern Ontario in the 1910s and 1920s, forest and prairie fires played a minor role in Canadian literature, and even less so in art. There was Alexander McLachlan's (1818–1896) poem "Fire in the Woods," in which a farmer sits down and cries after watching a forest fire tear through his property. And in a similarly themed poem, "The Forest Fire," Charles G.D. Roberts (1860–1943) writes about a father who saves his son from a fire that destroys their home in the forest, only to succumb after he rides his horse back, presumably to salvage what he could.

In the art world, there was Denis Gale's charming circa 1860 watercolour of a moose swimming across a lake towards a fire at nighttime, and Paul Kane's *A Prairie on Fire*. Neither painting evokes the sublime, albeit exaggerated, realism of American George Catlin's oil on canvas, *Prairie Meadows Burning* (1820), in

which ominous swirling clouds of black smoke are chasing Indigenous people on horseback across a burning prairie. As Kane expert I.S. MacLaren notes, Kane's prairie fire occupies so little of the canvas back at the horizon that it seems to "kindle decorously than rage alarmingly."[11]

1.3 *Prairie on Fire*, oil on canvas by Paul Kane painted circa 1849–56.

But no one except the much-maligned Susanna Moodie (1803–1885) came close to conveying the reality of fire. In her memoir, *Roughing It in the Bush*, she bears witness to what happened when the deliberate burning of fallow in the fields on a hot windy day got out of control. Her fires were based on real fire that tore through her farm in 1834. "I had not felt the least alarm up to this minute; I had never seen a fallow burnt, but I had heard of it as a thing of such common occurrence that I had never connected with it any idea of danger. Judge then, my surprise, my horror, when, on going to the back door, I saw that the fellow, to make sure of his work, had fired the field in fifty different places. Behind, before, on every side, we were surrounded by a wall of fire, burning furiously within a hundred yards of us, and cutting off all possibility of retreat; for could we have found an opening through the burning heaps, we could not have seen our way through the dense canopy of smoke; and, buried as we were in the heart of the forest, no one could discover our situation till we were beyond the reach of help."[12]

Canadian poet, novelist, and literary critic Susan Glickman notes that for most of the early colonists, wolves, bears, long, hard winters, and the many other travails associated with living in the forest represented not so much a "wilderness of horrors but a space of opportunity." Trees were worthy adversaries that could be conquered by enormous labour.[13]

So, too were the many other challenges faced in settling in the wilderness.

A shotgun could dispatch a bear, and bounties could temporarily drive down wolf and coyote populations. Poisoned bait could kill "chicken hawks" that opportunistically preyed on farmyard fowl. The chill of cold winters could be warded off with beaver fur coats, hats, and mittens, a Hudson's Bay blanket that could be fashioned into a robe, and a warm crackling fire in a wood stove. The buggy bog or fen could be drained. Forest fires, however, were more formidable – unconquerable in those early days when there were no trained firefighting brigades on the ground to fight them – laying waste to all that hard-earned labour with astonishing power and speed and brute force.

A wildfire was the one wilderness of horror that most artists didn't have the desire or the experience to describe. The chaos of colours and the otherworldly tangle of scorched prairie, charred timber, and spongy bogs and fens that often stopped a wildfire in its tracks did not fit into the schemata of the landscape that many learned about and incorporated into their work. Fire had no place

I.4 *Prairie Meadows Burning*, oil on canvas by George Caitlin, 1832.

1.5 *Upper Part of the Mackenzies. Woods on Fire*, watercolour on wove paper by George Back, 1825.

in "the milk and honey, late-Victorian, God and Maple Tree" romanticism that F.R. Scott scorned in the work of poets like Bliss Carman (1861–1929).[14]

Fire, for the most part, existed outside the knowable because there was no science to explain it, no long-term memory of it, and because most people who sketched, painted, or wrote fiction dwelled in big cities in central Canada that rarely experienced fires of note. It was as peculiar as the tendency in the eighteenth and nineteenth centuries for artists to favour clear skies and fair weather over views obscured by cloud and fog. The joke among art historians is that fog did not exist until Joseph Turner and Caspar David Friederich came along. Even then, the fog effect in their masterworks was limited so as not to divert too much attention from boats and people in the foreground.

And so it was with wildfire in Canada.

Not until the Group of Seven painters – Tom Thomson in particular– and Margaret Atwood's novel *Surfacing* (1972) and "The Two Fires," a poem she included in *The Journals of Susanna Moodie* (1970), did viewers of art and readers of Canadian fiction get a sense that most of Canada is unforgivingly boreal: fiery, spongy, buggy, running with wolves and bears, swimming with beavers, and snapping with turtles and rattlers denning in peat. In *Surfacing*, set in an unnamed forest of Quebec, nature presents itself as something that does not disapprove or approve, has nothing to say except for the "fact of itself," which is what wildfire is all about. It's quite simply a chemical reaction, a propulsive

I.6 *Fire-Swept Hills*, oil on composite wood-pulp board by Tom Thomson, 1915.

oxidation of hydrocarbons that is shaped by terrain, weather, climate, and the nature of the combustible material around it.

It's no surprise, then, that the first and still the only book on the history of wildfire in Canada was written by Stephen J. Pyne, the prolific Arizona State University historian who was literally begged, funded, and accommodated by several Canadian institutions to do what no Canadian historian would do at the time: researching and writing *Awful Splendour: A Fire History of Canada*, published in 2007. *Awful Splendour* is exhaustingly comprehensive and "structurally complex," as even Pyne concedes, in its chronicling and analysis of the convoluted relationship that Canadians had with fire. Totalling more than five hundred pages, the book takes a sometimes sympathetic, but ultimately harsh view of the country's response to this force majeure. There was, Pyne concludes, no real cultural engagement across scholarship, the arts, and politics. Canada, he said, "was more big than important. It was a partner, not a leader, a supplier of goods and techniques, not an originator of new ways to live on a fire-prone land."[15]

Pyne was right for the reasons mentioned above, and for many others. Canada did not have a Henry David Thoreau who, while embracing fire as a regenerative force in the 1850s, set off a very public debate about forest succession. Nor did it produce a mind like that of George Perkins Marsh whose 1864 book, *Man and Nature; or, Physical Geography as Modified by Human Action*, helped kickstart the conservation movement in America. There were in Canada Clifford Sifton and geologist Robert Bell, who championed the idea of forest conservation, but there was no Canadian as impactful as John Muir and Aldo Leopold were south of the border when it came to writing about fire. In national magazines such as *The Atlantic*, Muir routinely skewered lumber and railroad companies for allowing fires they set while clearing timber to run into the woods.

Canada got a good start when J.H. Morgan was appointed to conduct a commission on forest conservation in 1884. But his recommendations and those of his insightful contemporary, Robert Bell, were largely ignored and quickly forgotten. South of the border, Franklin B. Hough's national investigation of wildland fire two years earlier, on the other hand, was a revelation that endured in a succession of events that followed. It was Hough who set the stage for Gifford Pinchot's studies of wildfire in 1898 and the policy of fire suppression that Pinchot put into effect in 1905 when he headed up the US Forest Service.

Pinchot would have found no comfortable place in Canada's bureaucracy. Speaking directly to the press in the interest of the public good, as Pinchot had so often done in the US to further a national forest and wildfire strategy, was not the British/Canadian way. Bureaucrats in Canada were more accustomed to practicing procedural correctness than to stepping out of bounds for the benefit of public interest. This may have been the reason why Ernest Finlayson, the man who headed up the Dominion Forest Service from 1924 to 1936, went into a deep depression before disappearing without a trace after the federal government slashed his budget to almost nothing, forcing the closure of offices across the country and the loss or destruction of decades of records. Finlayson is mentioned ever so briefly in *The Globe* just three times during his long career. Pinchot, the American, made headlines in that paper more than thirty times.

Like Creighton, Lower, and other historians, Pyne assumes that the history of Canada – the struggle to adequately manage and live with wildfire in this case – can be best explained by the fractured east-west relationships between the provinces and the federal government, and the British institutional designs that made governing such a big, bilingual country problematic. Again, there is much truth in this theory. French Canadian nationalism and the provincial desire and legal right to control natural resources made it difficult for the federal government, impoverished and uninterested as it tended to be in investing in forest conservation, to assert a national forest and wildfire strategy outside of the national parks. Only when the provinces came calling for help during big fires years later did federal programs such as a national air tanker fleet and the Canadian Interagency Forest Fire Centre come into play in a meaningful way.

I.7 Elihu Stewart (1844–1935), first director of the Dominion Forest Service. Photo by William James Topley.

The military was also there to help when things got desperate, as they did in in 1989 when the premier of Manitoba called Canadian minister of defence Bill McKnight in the middle of the night for help with the biggest fire season in that province's settled history.

But as much as the fire history of Canada is, as Pyne says, "a confederation of narratives with separate histories,"[16] a case can be made, as he occasionally and implicitly does in his book, for a more fluid north-south chronicling in which Canada and the United States dealt with wildfire and forest conservation co-operatively, and sometimes competitively, beginning in 1882 when concerns for forest health on both sides of the border were first addressed in meetings of the American Forest Congress in Cincinnati and Montreal.

One of the first things that Elihu Stewart did when, after serving as chief inspector of forestry and timber for six years, he became the first superintendent of Forestry in 1905 was to write a letter to Pinchot asking him for advice, which was gladly given. Stewart recognized US expertise in forestry and tried to copy what Pinchot achieved under the presidency of Theodore Roosevelt.

What followed in Canada was a meeting of minds at another conference in Ottawa in 1906, and in 1907, the establishment at the University of Toronto of a school of forestry modelled after those at Cornell and Yale. Bernhard Fernow, a Prussian-born forester who was third chief of the US Division of Forestry, and dean of the school at Cornell, headed up the school at the University of Toronto from 1907 to 1919. Fernow sat on the Commission of Conservation (COC),

1.8 Bernhard Fernow (1851–1923), dean of Forestry, University of Toronto.

which was created a year after Roosevelt established the National Conservation Commission in the United States. The COC provided the government with up-to-date scientific advice on forests, lands, minerals, fisheries, game and fur-bearing animals, and water. In that role, Fernow recruited James H. White, the first forester to graduate from the University of Toronto, to take on the job of chief forester at the Commission of Conservation.

Canada followed the United States' lead in expanding forest reserves across the country. But it was the first country to create a national parks branch in 1911 that had responsibility for fire management. In 1932, Glacier and Waterton Lakes national parks became the first International Peace Park. When the Boundary Creek Fire burned along the border of Montana and Alberta three years later, firefighters from Canada and the US joined forces with men from the Civilian Conservation Corps, a national relief program that was created in 1932 to put Americans back to work. That set the stage for a formal wildfire-specific agreement in 1983. Both agencies collaborated on the 2015 Waterton Lakes Fire, the 2017 Kenow Wildfire, and the 2018 Boundary Wildfire. The Glacier National Park superintendent delegated authority to Parks Canada to manage the 2015 Waterton Lakes Fire although it was entirely on the US side of the border. Similarly, a Type 1 US Incident Management Team managed the 2018 Boundary Wildfire when it crossed into Canada.

It's not clear who initiated it, but sometime around 1937, Ontario and Minnesota began coordinating flights along their borders to maximize wildfire-detection coverage. The border flights continue to this day. Two years later, Canada created its own version of a Civilian Conservation Corps – unsuccessfully as it turned out – to help fight fires and plant trees as the United States had done in 1932. A year after the Bar Harbour Fire famously burned celebrity and millionaire homes in Maine in 1947, many of the east coast states joined forces with two Canadian provinces to form what is now a more expansive Northeastern Forest Fire Protection Commission, commonly known as the Northeast Compact. It is one of eight compacts that eventually came into play to allow states and provinces to share personnel, training, expertise, and equipment to minimize the firefighting burden during periods of high fire occurrence.

When the US invented "Smokey Bear," much of Canada adopted the famous bruin for its forest fire prevention campaigns. Alberta, on the other hand, opted for its own home-grown image, even though it ended up hiring the Walt Disney Studio to come up with "Bertie Beaver." Canada did not need to follow the US lead in shamefully evicting Indigenous people from national parks, partly because of their penchant for lighting fires. It already had a lot of experience crushing them with disease and false promises, and forcing them onto reserves.

Looking back now one can see, as Pyne points out, notable differences between the two countries. In the US, the Forest Service grew and prospered fighting fires and researching them. In Canada, the Dominion Forestry Branch and the Dominion Forest Service and the Canadian Forest Service (not to mention the many other iterations that were regularly shunted from one department to another) struggled to remain relevant, even to this day, partly because it ceased to be a firefighting agency long ago. Canada took fire behaviour research to the field with test fires while the Americans did most of their research in labs. Whether that made the United States a leader in fire science is debatable. There is no credible evidence that Americans were, or are, better at fighting blazes or living with them because of what came out of the laboratory. Canada may have lagged behind more often than it should have, but it always caught up, and surpassed its partner in some cases – the invention of the amphibious water bomber being one example. Neither country sufficiently appreciated that aggressive fire suppression would eventually fail.

It was these field experiments in Canada that shed light on the "nuclear winter" theory which envisioned smoke from hundreds of enormous fireballs caused by nuclear explosions lethally cooling the climate. The experiments also gave credence to the idea that pryoCbs – fire-driven thunderstorms –

had enough energy in them to rise above the troposphere, which contains half of the Earth's atmosphere and where weather occurs. Those insights came from the collaborative work by Mike Fromm of the US Naval Research Lab in Washington, DC, Brian Stocks of the Canadian Forest Service in Sault Ste Marie, Ontario, and Rene Servranckx of the Canadian Meteorological Centre in Dorval, Quebec, in the early 2000s.

Canada did not embrace "natural prescribed fire" or let lightning-ignited fires burn themselves out as a management strategy as early and aggressively as the US did, partly because of politics but also because some of the lightning-caused burns in US parks and wilderness areas resulted in catastrophic runaway fires that ultimately burned over massive areas, forced large-scale evacuations, destroyed many structures, and cost millions of dollars to contain. The 1988 Yellowstone Fires, according to Cliff White, the architect of Parks Canada's wildfire management strategy, put the fear of God into the agency. The unpredictable outcomes of such "natural fires" didn't stop Parks Canada from burning, he told me during a field trip in Banff; they did, however, stimulate Canadians to follow a cautious "planned-ignition" program in which managers focused on the ecological outcomes of burns rather than ignition sources.

White is another example of that fluid north-south relationship. He was schooled in the art of managing fire in the US, as were a long line of experts dating back to the days when Edmund Zavitz, the man who oversaw Canada's first attempt at reforestation in 1905, graduated from the forestry school at Michigan State, and Abraham Knechtel graduated from the Forestry School at Cornell before joining the Dominion Forestry Branch in 1910. Years before Aldo Leopold came along, Knechtel was preaching about the "other blessings of the forest" and their capacity to "feed springs, prevent floods, hinder erosion, shelter from storms, give health and recreation, protect game and fish, and give the country aesthetic features."[17]

Notable among the other graduates were Willis Norman Millar and Harvey Reginald MacMillan. Millar had an impressive career with the US Forest Service before joining his former schoolmate at the Dominion Forestry Branch in 1911. He became chief inspector of forest reserves the following year, when MacMillan moved to a new job in British Columbia. Millar's experience in the US undoubtedly shaped the way in which he made improvements to fire control in Canada.[18] He conducted extensive field trips on horseback and made sure that trails connecting the new ranger stations were well maintained. Lookouts and telephone lines were put in place upon his recommendation, and a message code was developed so that even the most poorly trained signaller could report the location of a fire.

Contrary to what Pyne concludes, many Canadians really were engaged, asking over and over again the question that Americans like John Wesley Powell first posed in 1878: "Can these forests be saved from fire?"[19] The evidence is not in history books, poems, novels, or the biographies of prime ministers and politicians such as Clifford Sifton, the former cabinet minister who headed up the Commission of Conservation for more than a decade. Nor is there much in the records of the Dominion Forestry Branch, or the Dominion Forest Service or Canadian Forest Service that followed in a long history that began with Elihu Stewart being appointed superintendent of what was originally the Forestry Branch in 1901. But I found ample evidence of this engagement in newspapers and magazines in Canada, the United States, and Britain, which followed the situation in North America very closely. The editors of newspapers and magazines, big and small, were obsessed with fire, the damage it did to property and forests, and in the number of people it killed and displaced. It was a crime story that kept on giving and a science story that bewildered as much as it dismayed. Mysterious, malevolent, and ready to pounce when people least expected, fire never failed to shock and appall, as it did in 1870 when a headline in the *Perth Courier* announced in a headline, "The Whole of Central Canada in One Mass of Flame."[20]

This relentless coverage of wildfire led to several royal commissions, many reports for the Commission of Conservation, heated public and political debates over matters such as using airplanes and parachuters to manage fire, and whether Indigenous people really were responsible for lighting so many of them. It wasn't just Canadian newspapers that informed readers about fire. Wire services such as the Associated Press disseminated news about fire from both sides of the border. The *New York Times* and the *Washington Post* paid as much attention to big conflagrations such as those that burned in Canada in 1870 as the *The Globe* and *Toronto Daily Star* did to those in many parts of the US the following year. Even the *Times* of London and the *Illustrated London News* were engaged in what was going on in the forests of North America. There was really no border when newspapers made the decision to run a story about fire. I'm not the first to make use of newspapers to revise and fill in the details of fire history in North America to demonstrate that the response to wildfire was as a much a north-south cultural and political phenomenon as it was an east-west one. American journalists such as Timothy Egan, author of *The Big Burn*, have been doing this for some time.[21]

Canadians are beginning to catch up now that Creighton, Lower, and other historians of their generation are no longer in vogue in the writing and teaching of Canadian history. (Creighton's death in 1979 made national news. Had he lived on, he would have been mortified to see how race, class, gender, religion, environment, and even beer now dominate the study of Canadian history.) Historian Alan MacEachern livens up *The Miramichi Fire: A History* (2020) with all manner of wonderful stories – true, false, and sketchy – that he gleaned from local newspapers like the *Quebec City Gazette*, which gleefully mocked a rival paper for suggesting that the fire of 1825, the biggest in North American history, was the product of spontaneous combustion, and large newspapers like the *New York Times* which gave credence to the unfounded rumour that the blaze was set by "two special constables" bent on evicting French Canadians. MacEachern's sleuthing and his scanning of these papers demonstrated throughout many parts of his book that the Miramichi – a fire that Pyne described as "the first historic holocaust of the reclamation"– was part of larger complex of fires that extended south into Maine. It was also one that engaged Maritimers for decades as part of the school curriculum.

Reading dozens of newspapers in Canada, the United States, and Great Britain from the 1850s to the present, I found many extraordinary stories of wildfire that have never been told, details of historic fires that have not yet come to light, fires that have not found their way into scholarly works, and informed discussions on how best to prevent fire. Some of the first-hand accounts were so artfully written that they could have been turned into books. Newspapers and magazines, I learned, were more than happy to demonize fire on behalf of the US and Canadian governments so long as they were paid to run their advertisements. They were also suitably indignant when governments failed to adequately deal with the challenge of fire.

So much has happened since Pyne's book was published that has reshaped the fire history of Canada as well as the United States. In addition to MacEachern's excellent history of Miramichi, we have Cordy Tymstra's book *The Chinchaga Firestorm* (2015), about a 1950 fire that burned so big in northern Alberta, it sent a pall of smoke to New York state turning day into night and caused many people to believe that the end of the world was at hand. As smoke drifted across the ocean into the air space over Great Britain and northern Europe, the sun and moon turned blue, causing a run on a bank in at least one case. A growing number of historians such as MacEachern, John Wadland (*Ernest Thompson Seton*), Tina Loo (*States of Nature*), and John Sandlos (*Hunters at the Margin*) have written the kind of environmental histories that Pyne, A. Donald Worster, B. William Cronon, C. Alfred Crosby, and D. Carolyn Merchant previously wrote in the US. Environmental history in the US got a kickstart at a roundtable in 1990 when this group of historians urged their colleagues to give voice to the "wisdom of nature" to assess how humans have altered the landscape for their own selfish purposes.[22]

The most important development since Pyne wrote his book is the way that fire has responded to climate change. The wildfire situation in North America, as in most of the world, has become much more volatile, killing more people, destroying more property, and forcing evacuations at rates that could not have been imagined in the early 2000s. So-called "bad fires" are quickly overshadowing the good ones that help regenerate forests and make them more resilient to disease, drought, and fire. And fire seasons are lasting weeks and in some cases months longer than they did before the turn of the century. Continuing to suppress fire as vigorously as we have done since around 1910 has, in many cases, shifted the balance away from young, mixed forests that are resilient to older forests that are much more likely to ignite and burn intensely. The draining of wetlands that can stop or slow a fire has also made things worse. In addition, whereas weather used to predict what a fire would do, increasingly, fires are determining what the weather will do. Warming temperatures are producing more lightning and igniting blazes at higher rates, spreading with speeds and burning with intensities that are rare or unprecedented. They are also spawning fire tornados and pyroCbs – fire-induced thunderstorms that create their own lightning. Hot, dry places such as California used to benefit from rain. But now rain, in some cases, is expediting the growth of plants that will dry up and end up as fuel for fire. Events such as these were so rare before the turn of the century that most scientists considered them to be inconsequential. It's not like we haven't been warned. The fire scholarship in Canada – that Pyne suggests wasn't there – foresaw an increase in fire and fire intensities in a warming climate as early as 1991, when Charlie Van Wagner and Mike Flannigan published a study on how climate change might impact wildfire in the boreal forest of Canada.[23] In an interview with me, Flannigan admits that neither he, nor his colleagues, envisioned that the world would burn as big and chaotically as it is burning these days in every part of the world outside of Antarctica.

North Americans are no longer in denial to the degree they were in the past because the powerfully "bad" conflagrations that produce fire tornados and pyroCbs are impossible to ignore when they come as close as they do now to people's doorsteps. Big cities such as Vancouver, Toronto, New York, and Montreal may not be directly in harm's way, but fires are sending them smoke that can turn day into night and make breathing unpleasant and unhealthy. Denial, however, continues to persist in government circles, as politicians pay out billions in compensation while balking at investing more generously in fire science, defensible spaces, and ways in which the impact of fire can be mitigated. The situation is improving, as is addressing the larger issue of climate change, but not nearly as fast as it needs to in order to keep pace with what's happening on the landscape.

Lessons learned from the past can teach a great deal about the present and how we might better live with fire in the future. What the future is beginning to tell us is that business as usual will not be successful.

1

PRELUDE TO THE DARK DAYS AT NOON

On the morning of 19 May 1780, just a week after their biggest defeat during the American Revolution, General George Washington and his army were at headquarters in Morristown, New Jersey. It was an inauspicious start to the day. The sun rose in the east with a reddish rather than yellow glow. Following a night when the full moon presented itself as pink instead of white, the future president took note of the strange weather in his diary. "Heavy and uncommon kind of clouds," he wrote, "dark and the same time a bright and reddish kind of light intermixed with them."[1]

Washington was not alone in sensing that something was amiss. In Weston, Massachusetts, Abigail Adams, wife of future president John Adams, expressed similar sentiments in a letter she wrote to friend and politician William Lovell from her home in Braintree, Massachusetts.

> We have had a strange Phenomenon in the Natural World. On fryday the 19 of May the Sun rose with a thick smoaky atmosphere, indicating dry weather which we had for ten days before. Soon after 8 oclock in morning the sun shut in and it rained half an hour, after that there arose Light Luminous clouds from the north west, the wind at south west. They gradually spread over the hemisphere till such a darkness took place as appears in a total Eclipse. By Eleven oclock candles were light up in every House, the cattle retired to the Barns, the fouls to roost and the frogs croaked ... I hope some of our Philosophical Geniousess will endeavour

> to investigate so unusual an appearance. It is matter of great consternation to many. It was the most solemn appearance my Eyes ever beheld but the Philosophical Eye can look through and trust the Ruler of the Sky.[2]

Judge Samuel Phillips Savage, a close friend of Samuel Adams and one of the founders of the Tea Party, noted in his diary that a "remarkable thick air" had permeated the skies for days. "The sun rises and sets very red," he wrote. At 9 a.m. on that fateful Black Friday, there "came on an appearance over the whole visible heavens … a light grassy hue, nearly the color of pale Cyder." The sky soon "attended with gloom resembling that of an Eclipse of the Sun."[3]

From the border of New Brunswick and Maine south to Connecticut, Rhode Island, and beyond, thousands of people responded with bewilderment, panic, terror, and despair as the unearthly sky grew darker and more ominous with each passing hour. Shopkeepers and homeowners lit candles so they could see what they were writing or reading. Outside, farmers abandoned the fields and wildlife behaved as if the light of day were done. "The birds have sung their evening songs. The fowls retired to roost," recalled Samuel Williams, clergyman, naturalist, and for a time the Hollis Professor of Mathematics and Natural Philosophy at Harvard College. "The cocks were crowing all around, as at break of day; objects could not be distinguished but at a very little distance; and everything bore the appearance of gloom at night."[4] It was an ominous sight for a scientist who was in the process of calculating the best place to observe a total eclipse of the sun, which was expected later that year. Williams wasn't sure what was going on as the rain that eventually fell "was found to have an uncommon appearance, being thick, dark and sooty … Our intelligence in this is not so particular as I could wish," he wrote.[5]

As midday quickly turned to night, citizens rushed to churches to pray and commiserate by candlelight. Clergyman Timothy Dwight observed that most of them believed that "the day of judgment was at hand." Sailors, according to Salem lawyer William Pynchon, weren't nearly as intimidated as they went "halooing and frolicking through the streets," mocking ladies as they scurried by. "Now you may take off your rolls and high caps and be [damned]," they yelled out.[6]

Lawmakers throughout New England adjourned early because many of them also believed the darkness was a sign that the end of the world was at hand. Abraham Davenport, a member of the Connecticut Legislature, urged his fellow lawmakers to stay put. "I am against adjournment," he declared. "The day of judgment is either approaching, or it is not. If it is not, there is no cause for adjournment; if it is, I choose to be found doing my duty. I wish therefore that candles may be brought."[7]

For more than two centuries, scientists and philosophers puzzled over the source of this terrifying darkness. Those who were deeply religious, and most were back then, speculated that it was a warning from God, as prophesized in Biblical passages such as Joel 2:31: "The sun shall be turned into darkness, and the moon into blood, before the great and the terrible day of the Lord come." An editor of *The Williams Quarterly Journal* at Williams University in Virginia noted that even "after they saw the regular succession of the day and night restored, and found that the trump was not sounded, nor the dead raised, they still believed it to have prognosticated some dire event, some judgment for the nation's sins."[8]

Although some smelled smoke that day, many dismissed the possibility of a distant wildfire as too "simple and absurd,"[9] even as others found rainwater collected in tubs with "light scum upon it, which, upon being rubbed between the thumb and finger, seemed to resemble the black ashes of burnt leaves."[10] Ezra Stiles, the president of Yale College in New Haven, offered one of the more plausible explanations in noting that "the woods about Ticonderoga [in New York] and eastward over to New Hampshire and westward into New York and the Jerseys were all on fire for a week before this Darkness and the smoke in the wilderness almost to suffocation. No rain since last fall, the woods excessively dry ... Such a profusion of settlers pushing back into the wilderness were everywhere clearing land and burning brush. This set the forests afire far beyond intention, so as to burn houses and fences ... The woods burned extensively for a week before the nineteenth of May and the wind all the while northerly."[11]

In reconstructing that "Dark Day" in 1780, a group of contemporary scientists have estimated that the pall spread 290 kilometres, seven and a half hours from Portland, Maine, to southwestern Massachusetts, at a speed of forty kilometres per hour. University of Missouri scientist Erin R. McMurry and colleagues at the US Forest Service used tree ring data to link the possible cause to forest fires that burned in several locations, including the area that is now Algonquin Provincial Park in Ontario, western Maryland, western Virginia, the Missouri Ozark highlands, the Boston Mountains of Arkansas, and the Boundary Waters Canoe Area Wilderness of northern Minnesota. The year 1780, they point out, was the greatest year of fire recorded in eastern North America before 1850.[12]

The events of 1780 were prophetic of the role that climate, weather, forests, fire, and the growing number of people on the landscape would play in shaping how governments and individuals would think of risk and loss. There had been a few "dark days" before and there would be many more on both sides of the border in the decades that followed. But as each one passed, it seemed that no one was the wiser for it.

"Business as usual," the pre-industrial approach to business practices and lifestyle decisions, ensured at least a reliable form of subsistence, famine and plague notwithstanding. In the New World, that outlook eventually gave way to riskier but more profitable ventures such as trapping beavers for the lucrative fur trade, indiscriminately slaughtering bison for hides and leather, draining wetlands, burning and clearcutting forests for agriculture, mines, timber, pulp and paper, and for energy developments. Cities, homes, and railway and utilities lines were built in or near forests that relied on fire to regenerate them as they grew old and increasingly vulnerable to disease.

The fires of 1780 did not hurt anyone or destroy any property, as far as we know. What they did do, however, was foreshadow the beginning of an increasingly unstable and rapidly warming climate as a result of industrial activity.[13] Most people in North America and, in fact, most people throughout the world, either ignored or didn't understand or refused to consider any of this in predicting future losses and gains. A roll of the dice took place in which the odds of societies winning the battle with large and increasingly intense wildfires were stacked against them. Even as progressive newspapers sounded the alarm over and over again, laissez-faire governments were willfully slow to step in, content in the belief that free markets could solve all unforeseen problems.

As it eventually became evident that free markets would not, or could not, sort this out, there was a blind assumption that systematic prevention and suppression of wildfires would suffice. No one dreamed that fires would eventually create thunderstorms that would trigger blazes thirty kilometres away, or that they would smoulder under the snow for the winter before coming back to life like zombies in springtime. Nor did anyone imagine that cities would burn, that utilities and water treatment plants would fail, and that a fire season like the one that ripped through Australia in 2019–20 would kill more than a billion animals. The "new firsts" that are now so common were beyond everyone's grasp.

Underlying this game of Russian roulette was a form of environmental racism in which Indigenous people – who understood the important role that fire played in making forests and grasslands more productive and resilient – were seen as part of the problem. Rather than embracing the light burning strategy with which Indigenous people had managed forests for hunting and for growing food and harvesting herbs, wild berries, and root vegetables, governments systematically removed Indigenous people from public lands and prohibited them from igniting fires, while allowing settlers to burn as they pleased and to behave in ways that invited fire to come.

2

THE FIRE TRIANGLE

Oxygen, fuel, and heat, the three components of the so-called "Fire Triangle," are all that are needed to start a blaze. No one knows exactly where jolts of energy hot enough to light the "Dark Day" fires of 1780 came from. There were no active volcanoes erupting or giant meteorites crashing to earth that year.[1] Fires lit by farmers or Indigenous people could have been responsible. Lightning strikes are the only other explanation. The Algonquin area experiences lightning fires a great deal – up to ten for every 400,000 hectares each year.[2] High-intensity fires burned in the Boundary Waters Canoe Area Wilderness of Minnesota about every four years before settlers moved into the region.[3] West Virginia is famous for its thunderstorms, and Indigenous people in the Ozark Highlands region and Arkansas River valley were managing the landscape with controlled burns in 1780 when a near decade-long drought began.[4]

No one knows how much wildfire was caused by lightning before the first Europeans settled in North America. Presumably many of these loomed large because there was no one there with the resources to put them out. What we do know from charcoal layers in sediments and methane concentrations in ice cores is that global wildfire activity fluctuated dramatically over time, and that there was a distinct lull in burning from about 1600 to 1750.[5] This lull occurred during the Little Ice Age (1305 to 1825), when temperatures in eastern North America and western Europe dropped by as much as two degrees Celsius. That's significant, because lightning strikes rise or decline by about 12 per cent with each degree of temperature gain or loss.

2.1 The early settlers of North America came at a time known as "the Little Ice Age" when there was less fire, 1817–62. Photo by Edward Struzik.

That cooler climate would have made it difficult for big fires to burn in eastern North America as they did with regularity in the west and in the boreal forest regions of the north, where highly combustible conifers such as tamarack, black and white spruce, and lodgepole pine dominate many ecosystems. Some of these tree species are born to burn because the heat of fire is required to get the cones to open up and release their seeds. If there were no fires, forests like these would not have existed over such large areas of the continent. Other species require fire to open up areas where their seeds are more likely to take root.

In the damper fens, bogs, marshes, and swamps along the east coast, the moisture would have made it even more difficult for fire to spread. The wetlands that dominated much of the landscape from New York and New Jersey south to

North and South Carolina, Georgia, and Florida would have eventually stopped big conflagrations from spreading before they were systematically drained in the nineteenth and twentieth centuries.

If there were large fires burning often in eastern North America when the first colonists arrived from Britain, Poland, Holland, and Germany, there is little evidence of it in public records prior to 1780. Nor is there anything in tree rings or lake cores to suggest that the burning that did occur was frequent, intense, and expansive. In fact, large fires were almost absent from most of Maine before that time.[6] The situation was different in the west, where it was warmer and drier. Again, no one knows for sure because the record for prehistoric fire regimes is patchy at best. However, there is evidence to suggest that fire activity on the Great Plains of North America was high between 1100 and 1650, and possibly even higher in the 1790s when traders and settlers began moving into the region.[7] That has been corroborated by a study published in 2021 that assembled and analyzed 795 reports of fires in present-day Illinois, Indiana, Iowa, Minnesota, Missouri, North Dakota, Ohio, and Wisconsin.[8] Presumably many of the border state fires had crossed over and migrated in from Canada.

There was, and still is, a myth posited by naturalists such as Henry David Thoreau that pre-Columbian ecosystems were pristine ones in which Indigenous people lived in complete harmony with nature, neither manipulating nor disturbing it for any reason other than subsistence. “Thus a man shall lead his life away from here on the edge of the wilderness, in Indian Millinocket stream, in a new world, far in the dark of a continent, and have a flute to play at evening here, while his strains echo to the stars, amid the howling of wolves; shall live, as it were, in the primitive age of the world, a primitive man.”[9] The reality is that most Indigenous people were masters, rather than slaves, of fire. They burned lightly to clear trees and brush for planting crops, to promote the growth of wild berries, medicinal herbs, and root vegetables, to encourage large animals to graze on fresh grass and tree shoots, or to drive animals such as bison and deer into lanes or tightly knit groups where they could be more easily hunted.[10]

The notion that Indigenous people were responsible for most of the fires in pre-Columbian America is a subject of debate that will likely go on for some time. But one thing is certain. Evidence from archeological sites and from the narratives of explorers and fur traders make it clear that Indigenous burning was common by the time Europeans arrived in North America. Most early newcomers were puzzled by this practice. When the first Jamestown colonists landed in 1607 and saw “great smokes of fire” rising from the forest, they assumed that the inhabitants were summoning other tribes to battle. They soon discovered that they were clearing the land in order to grow corn, squash, tobacco, and other plants just as the settlers themselves did to clear farm fields back home in Great Britain.

2.2 *Blackfeet Burning Crow Buffalo Range*, watercolour by Charles Marion Russell, 1905.

This curiosity eventually turned to consternation as the agricultural ambitions of the newcomers – cutting down trees, burning brush, and draining swamps to grow tobacco, corn, and cotton, to build towns and to raise livestock – clashed with Indigenous burning. In 1730, British colonist William Bohannon stepped into the Orange County courthouse in Virginia to make an oath "that about twenty-six Saponey Indians who had inhabited Colonel Spotswood's land in Fox's Neck go about and do great mischief by setting fire to the woods."[11] Bohannon acknowledged that Spotswood, the former lieutenant governor, may have allowed the Saponey to camp and to hunt on his land. But he was clearly incensed by this wanton act of burning a perfectly good forest, as well as threatening the pigs he allowed to root in it. The practice of fencing in livestock would come much later.

There are no records to indicate how the authorities responded to his oath. But as years went by, cultural and economic clashes over Indigenous burning intensified pretty much everywhere Europeans went. It was fine for settlers to burn the land for the export of potash and for domestic agricultural purposes, and to grow tobacco in quantities that depleted the soil, but it was viewed as irresponsible and borderline criminal if Indigenous people did the same things. Potash, potassium carbonate that comes from the residue of burned wood, was being shipped off to England in enormous quantities by the time of the American Revolution.

Some Europeans did see Indigenous burning for what it was. Juan Crespi, a friar who served as chaplain on a Spanish expedition to California's central coast in 1769, is most famous for giving the city of Los Angeles its name. During an overland expedition from the Bay area to present day San Diego, he observed that the Quiroste people regularly burned the meadowlands "for a better yield of the grass seeds that they eat."[12] In further describing what he saw, he noted: "We ascended a little hill and entered upon some mesas covered with dry grass, in parts burned by the heathen for the purpose of hunting hares and rabbits which live there in abundance."[13] And decades later, in 1804 and 1805, members of the Lewis and Clark expedition made similar observations when they passed through several recent or active fires on their westward trek. The purpose of the expedition was to explore the Louisiana Purchase, 2,100,000 square kilometres of land that the US had bought from France for $15 million. President Thomas Jefferson had appointed Meriwether Lewis as commander of the expedition and instructed him to establish friendly relations with Indigenous people so that the US could expand the fur trade and find the Northwest Passage, which was thought to be a short cut to the Orient. The Americans had no realistic idea of what to expect. Jefferson was not alone in believing that Lewis and William Clark would encounter woolly mammoths and possibly a Welsh-speaking tribe that, according to folklore, were descendants of an 1170 expedition to the Americas financed by a Welsh prince.

Neither mammoths nor Welsh-speaking Indigenous people were found. But expedition members did observe raging wildfires that made parts of the journey even more challenging than they would have been otherwise. Seven of the ten blazes the expedition encountered were likely ignited by Indigenous people.[14] Lewis suggested that in at least one case, Indigenous people "had set the plain on fire to alarm the more distant natives and fled themselves further into the interior of the mountains."[15] Clark, on the other hand, speculated that some of fires were intended to improve browsing opportunities for the horses the Indians rode and for the buffalo and deer they hunted. "The plains are on fire in view of the fort on both sides of the river," he wrote on 30 March 1805. "It is said to be common for the Indians to burn the plains near their villages every spring for the benefit of the horse and to induce the Buffalow to come near them."[16]

In the process of mapping Canada and the Kootenai region of the American northwest in the early 1800s, fur trader David Thompson heard rumours about mammoths dwelling in the mountains, but doubted their existence. Some of the Métis and Indigenous people who accompanied him, however, held "a strong

2.3 Explorer David Thompson (1770–1857).

belief that the haunt of the Mammoth, is about the defile," he wrote. "I questioned several, none could positively say, they had seen him, but their belief I found firm and not to be shaken."[17]

Thompson was a keen observer of Indigenous people's practices and an admirer of their way of living off the land. His fifty-seven-year marriage to Charlotte Small, a Métis woman, is a testament to that. Unlike Lewis and Clark, Thompson understood that the fires Indigenous people lit were not meant to destroy but to bring new life to the prairies and the forest – a subtle difference perhaps, but an important one. "The mercy of Providence has given a productive power to the roots of the grass of the Plains and of the Meadows, on which the fire has no effect," he wrote. "The fire passes in flame and smoke, what was a lovely green is now a deep black; the rain descends, and this odious colour disappears, and is replaced by a still brighter green; if these grasses had not this wonderful productive power on which fire has no effect, these Great Plains would, many centuries ago, have been without Man, Bird or Beast.[18]

There was a science to it which many Indigenous scholars today refer to as "traditional ecological knowledge," or TEK. Indigenous people tended to burn in spring and fall when the weather was cool, damp, and less likely to cause fires to spread over larger areas. On the other hand, some fires, like those on the prairies, were meant to burn big in order to keep the bison healthy. Fur trader Alexander Henry reported seeing one such fire raze a large area near Pembina (North Dakota) in 1800, and another rage for twenty-six days in the same region during the fall of 1807. He did not appreciate what was being accomplished. "The sight was awful indeed," he wrote. "This fire was a disagreeable affair for us at this season on the year, for should it continue its progress all over the Country we shall be hard put to for provisions as there will be no Buffalo and nothing can stop its fury but Snow or Rain."[19]

John Sullivan was a member of the Palliser expedition that explored and surveyed western Canada between 1857 and 1860. He was just as contemptuous of this practice. "It is most lamentable to see so often masses of valuable timber destroyed, almost invariably with wonton [*sic*] carelessness and mischief. Unfortunately, the Indians have a most disastrous habit of setting the prairie on fire for the most trivial and worse than usual reasons."[20] Like most Americans, Sullivan laid little blame on settlers who ignored a Council of Assiniboia decree of 1832 that prohibited people from burning between the beginning of March and the first of December. Many of these resulted in runaway fires. When it was Indigenous people who were responsible, however, it was another story.

The equilibrium between the Indigenous need to hunt bison and the practice of burning to help keep bison populations stable gradually shifted away from the animal's favour as more and more Europeans occupied the Great Plains of North America. In the Assiniboine River of western Canada alone there were, at one point, twenty-one Hudson Bay Company posts in operation, all requiring meat. As these forts and the number of people who inhabited them multiplied along the route to Fort Edmonton and beyond, the need to exploit bison increased dramatically, as it did on the Great Plains of the United States where commercial hide-hunters operated year-round. In order to secure that source of supply in Canada, Indigenous hunters began using fire to drive bison as far away from trading posts as possible so that the traders, unable to hunt the animals themselves, would be forced to purchase the meat instead. Fire became a weapon of economic warfare.

Many of the fur traders reluctantly recognized this advantage that the Cree, Sioux, and Blackfoot had over them. "The Plains around us are all on fire," Duncan McGillivray wrote from Fort George in 1794. "We hear that the animals fly away in every direction to save themselves from the flames, an attempt which is often rendered abortive when the fire is cherished by a breeze of wind which drives it along with such fury that the fleetest horse can scarcely outrun it. The

Indians often make use of this method to frighten away the animals in order to enhance the value of their own provisions."[21]

The strategy of steering bison away from Hudson Bay posts drove George Simpson, the governor of the Hudson Bay Company, to distraction. Unlike Thompson, Simpson was an unapologetic racist who referred to the many Indigenous women he bedded as his "bits of brown."[22] He wanted nothing from Indigenous people beyond the furs, hides, and meat they brought in to trade. To see Indigenous people successfully use fire to try to starve them out or force them to pay must have been alarming for a man who saw Indigenous people as inferior to settlers in every way. "The Company's establishment from [Fort Garry] to the source of the Assiniboine have even been at times in a state of Famine ... The failure of the buffalo may be attributed to two causes, one is that the plain Indians finding the coalition had taken place conceived that the sole object was as they express it 'to render them pitiful', and by way of having revenge determined on Starving the Traders by keeping the Buffalo off in the Summer and Fall, which was easily effected by obstructing them at their usual passes to the Northward, setting fire to the Plains."[23]

Between the 1790s, when the first Europeans began migrating west, and the 1880s, when the last of the wild bison herds had all but disappeared, scarcely a year went by without a major fire on the prairie landscape. Many no doubt loomed larger than they might ordinarily have if beavers had not been extirpated by Indigenous trappers who sold their fur to the Hudson's Bay Company and the Northwest Company. The loss of nearly 90 per cent of the beaver population in the United States and southern Canada during the nineteenth century greatly diminished the number of bogs, fens, and swamps that would otherwise have slowed or stopped the fires. Beavers are masters at engineering wetlands.

William Rannie was one of the few twentieth-century historians who saw wildfire as a seminal event. He painstakingly compiled more than 200 observations of fire from documents in the Hudson's Bay Company and Manitoba Archives and other historical sources. Most occurred in the vicinities of Red River Settlement, Pembina, Brandon House, and Fort Pelly.[24] For example, in May 1797, the Hudson Bay Company at Brandon House reported that the plains had been on fire for eight days. In 1801, Alexander Henry made note of "Terrible fires all over the plains." At Red River in 1821, Anglican minister John West described an "awful spectacle seen this evening through the whole of the northern, and western region." "Smoke suffocating," was the way it was described at Brandon House in 1828. Fort Pelly, a fur trade post operated by the Hudson's Bay Company at the elbow of the Assiniboine River in Saskatchewan, nearly burned down a year later.

In 1871 James Hargrave, the Hudson Bay Company factor at Red River, described what it was like to travel into the heart of a fire, suicidal as that seemed to be. "At first we caught sight of a fiery line faintly illuminating the far horizon on our front, Gradually, as we advanced, the line expanded into a crescent, extending to the right and left, and instead of one line, a vast number of blazing arcs broke on the sight. As we reached the heart of the conflagration, the entire horizon about us was luminous with low burning zones, whence the dark smoke curled aloft into the night."[25]

Wildfire was such a common event on the prairies and in the boreal forests of the west that Catholic missionaries such as Leon Doucet noted its occurrence with simple matter-of-factness. Stationed at Blackfoot Crossing in southern Alberta in October 1879, Doucet wrote that wolves were everywhere, and starving Cree arrived desperate in their search for bison. "I interpret for them … Norrish gives them a few provisions. They are going to Montana. Before leaving the Blackfoot set the prairie on fire."[26] Between 1868 and 1890, Doucet and his colleagues were constantly running away from or battling fires that threatened to burn their mission, but surveyor Edward Dawson, the man who built the first ranch in the Queenstown area of southern Alberta in 1888, wasn't as lucky as them. He lost his ranch to a prairie fire.

The Indigenous practice of burning was exceptionally sophisticated. The Pikunni people of Montana and the Piikani of Alberta, for example, used bison horns filled with mosses, softwoods, and hardwoods to carry fire from one camp to another. The vessel was covered in mud and clay and ventilated in a way to keep the fire smouldering. The Pikani could easily have started a new fire, according to Pikunni Blackfeet elder Marvin Weatherwax in a video interview he did with the US Forest Service. But the practice was more than just a strategy to keep the flame alive. It was a "spiritual" act that spoke to the longevity of the Pikunni and Piikani. Burning the land before leaving camp was a way of keeping the land "clean" so that food and pharmaceuticals could grow.[27]

The practice of burning and carrying the flame declined dramatically in the 1880s for several reasons. The extirpation of bison on the Great Plains of North America reduced the need to burn so that grass would grow to feed them. Persistent outbreaks of diseases such as smallpox, measles, scarlet fever, and influenza, to which Indigenous people had no immunity, dramatically reduced the number of people who could continue the practice. And because there were no buffalo left to hunt, Indigenous people didn't have the strength they needed to defend their territory from being taken over by farmers, railroad companies, miners, and timber harvesters. It was all too easy for Canadian and American authorities to herd most of them onto reserves where they were less of a threat. Taking their children away and putting them into far-off residential schools further broke their spirit, not to mention the respect for their cultural ways.

Estimates as to how many Indigenous people died as result of Old World diseases vary wildly, because no one knows the population of pre-Columbian

North America. Some anthropologists suggest the number was twelve million; some suggest it was as high as eighteen million.[28] Others claim it was just 900,000, which is still considerable. What we do know with more certainty is that there were only about 237,000 Indigenous people left on the continent by 1900.[29] By that time, anthropogenic fires – those caused by human behaviour – could no longer be legitimately blamed on them, though that didn't stop a lot of people from doing so. The settlers didn't want to face the reality that they were inflicting fire on themselves. Blindly taking risks for short-term profit was preferable to accepting the possibility that they might lose income and resources in the long run. Geological Survey of Canada scientist Robert Bell was one of the few people in the country who saw this unfolding in the 1870s and 1880s, when he lashed out at "white settlers" for burning so carelessly and indiscriminately. His cogent views on fire arose from extensive surveys as far north as Great Slave Lake in the Northwest Territories. Almost all the boreal forest, he observed, had burned at some time.

Bell appreciated Indigenous people's respect for fire, but he suggested that this respect was waning and in need of rehabilitation. Like John Wesley Powell, he believed fires needed to be suppressed by forest guardians, including Indigenous people, by whatever means necessary.

Bell's understanding of boreal forest ecology and fire behaviour would have served Prime Minister John A. Macdonald well when he suggested in 1871 that Canadian forests were being recklessly destroyed. The tragedy is that much of what Bell astutely advocated went unappreciated until 1976, when Canadian Forest Service scientist Marty Alexander came across many of the 200-plus papers he had written over a fifty-two-year career.

3

MORE DARK DAYS COMING

In 1845, Henry David Thoreau moved to the woods of Massachusetts because, as he famously stated, "I wished to live deliberately, to front only the essential facts of life, and see if I could not learn what it had to teach, and not, when I came to die, discover that I had not lived."[1] Like David Thompson, Thoreau came to appreciate that fire was a positive rather than negative force in nature. The revelation had come to him serendipitously when he accidently ignited a forest while fishing along the Sudbury River near his hometown in Concord, Massachusetts, the year before.

It had been an unusually hot, dry spring. By mid-afternoon, Thoreau and his friend, Edward Hoar, had caught as many fish as they had planned on catching. Before heading home, they decided to stop and make a chowder. Thoreau was well versed in the art of starting a fire. But on this day, a sudden gust of wind scattered embers onto the dry grass. Thoreau and Hoar did all they could but failed to stomp out the blaze. The fire "went leaping and crackling wildly and irreclaimably toward the wood," Thoreau recounted in his journal. Flames shot up to the tops of pine trees "as if they were powder, and squirrels and pigeons fled the forest as dense clouds of smoke billowed into the skies. We felt that we had no control over the demonic creature to which we had given birth."[2]

Realizing that the fire might burn the village, Hoar went off in their boat to warn its inhabitants. Thoreau stayed behind but failed to rally support from local farmers. He ended up on a cliffside, watching in despair as the fire burned the

3.1 Henry David Thoreau (1817–1862).

last pristine forest in the region. “I grieved with a grief that lasted longer and was more inconsolable than any of the proprietors,” he later wrote.

Villagers eventually arrived with hoes, shovels, and buckets. (There were no fire brigades back then.) By early evening, they had things under control. Forgiveness for the culprits, however, was not forthcoming. An editorial in the local newspaper blamed the blaze on the “thoughtlessness of two of our citizens.” For some time, many people refused to let Thoreau forget what he had done. They called him a “damned rascal,” a “flibbertigibbet,” and a “woods burner.”[3] These verbal attacks may have been what drove Thoreau to break with civilization and build a cabin in the woods at Walden Pond the following year.

In the weeks and months after the fire, Thoreau had a change of heart when he saw the grass growing and the trees suckering from the blackened ground. “I have set fire to forest, but I have done no wrong therein, and now it is as if the lightning had done it," he wrote. "These flames are but consuming their natural food."[4] Thoreau saw how fire regenerates "some of the noblest natural parks." When fire burns, he noted, it cleans the forest floor like a broom, and it becomes "inspiriting to walk amid the fresh green sprouts of grass and shrubbery pushing upward through the charred surface with more vigorous growth."[5]

It was a revelation, but not one that most people shared until George Perkins Marsh came along in 1864 and observed, in his book *Man on Nature*, how so much of the forest that burned in Maine in 1825 had recovered. Horace Greeley, Thoreau's friend, literary agent, and founder of the *New York Tribune*, was aghast at Thoreau's ideas about fire and forest succession. Thoreau's belief that the seeds of trees could survive being burned, and that wind, water birds, and quadrupeds brought them in when seeds were incinerated, ran counter to the scientific explanations of the day. Like many others, Greeley believed in "*abiogenesis*," the widely held theory that trees grow back as a result of spontaneous generation. According to the theory, plants simply come into being from non-living matter, or, as Loren Eiseley explains in her book *Darwin's Century: Evolution and the Men Who Discovered It*, "The unity of biological form is thus not the product of descent with notification, but rather a succession of creations linked only be an abstract unity in the mind of God."[6]

Thoreau, one of the first Americans to embrace Darwin's theory of natural selection, was clearly inspired by what he saw rising from the charred forest. In his later writings, he used fire imagery to express purification, wonder, and renewed friendship. Fire was not the demon that most people considered it to be, according to Thoreau. It was a physical force that underpinned the transcendental nature of the world in which he preferred to live. It was good rather than evil.

Thoreau was ahead of his time and far more insightful about the workings of nature than his contemporaries. Most early settlers came from Great Britain and other European countries where there were few if any wildfires of significance. A megafire, one that burns more than 40,500 hectares, was rare where they had come from, because forests there were pretty much gone. Just 2 per cent of Ireland's forests remained in the nineteenth century. In England, it was 5 per cent, maybe a little more. This is one of the reasons people were mining peat for fuel. Outside of Scandinavia and the more northerly parts of Europe, there were almost no trees left to cut down, not even in Iceland, which was once a reliable source of timber.

If, and when, fires burned out of control in the British Isles, it was usually because farmers burned brush to clear fields for planting or to condition the soil with charred wood and ash. Villagers extinguished them with hoes, shovels, buckets of water, and ditches that they dug around the burning grass and forest. More often than not, no more effort than that was needed. Moreover, the fens and bogs that were common throughout the British Isles stopped fires from spreading very far.

When the British began settling in North America, the Little Ice Age was still keeping things as cool and wet as it had been back home where rivers, lakes, and harbours routinely froze in winter and where rain and cold in summer often resulted in famine. Most people expected the weather to stay that way. In time, however, the Little Ice Age began petering out and a new reality unfolded. The higher temperatures and increased lightning that came with the warming temperatures set the stage for bigger fires like those of the Dark Day of 1780. The widespread draining of bogs, fens, and marshes, which began in earnest during Thoreau's time, meant that such blazes could spread far and wide.

The draining of wetlands was as industrious and thorough as the clearcutting of the forests. Henry Flagg French, an American farmer and, just as notably, assistant secretary treasurer, was one of a number of Americans who championed the idea of draining swamps, bogs, fens, and marshes on the east coast and in the remotest regions of the United States. His vision was a kind of Manifest Destiny for getting rid of the wetlands that were nature's way of containing fires. "In New England, we have determined to dry the springy hillsides, and so lengthen our seasons for labor; we have found too in the valleys and swamp, the soil which has been washed from on our mountains … On the prairies of the Great West, large tracts are found just a little too wet for the best crops of corn and wheat, and this inquiry is anxiously made; how can we be rid of this surplus water"[7]

Thirteen years before George Washington made mention of the "heavy and uncommon clouds" he saw on the Dark Day of 1780, Washington and others found a company whose main goal was to drain the Great Dismal Swamp on the Virginia–North Carolina border with the aim of selling the land as agricultural. The scheme appealed to wealthy Americans as well as the US Congress, which later invested in the drainage efforts. But drying out the vast wetland created conditions in which lightning could ignite the peat and allow it smoulder for weeks and months, as it continues to do to this day.

High-intensity fires, like the ones that burned in 1780, were not a concern because they occurred in places where few settlers lived. Indigenous people in the wooded area of the Algonquin region and the Boundary Canoe area of northern Minnesota simply packed up and paddled to safety in their canoes. Those in the Ozark highlands fled on foot or horse to familiar refuges. But as the drying forests in eastern North America began to fill up rapidly with people bent on draining wetlands and clearing forest for timber and agricultural purposes, the stage was set for a catastrophic fire like the one that burned in New Brunswick and Maine in 1825.

Up until 1800, Northumberland County along the Miramichi River was almost exclusively the domain of the Mi'gmaq. By the time of the Miramichi Fire in 1825, 8,500 people had settled there to farm and to work for the timber

companies. Like Indigenous people, European Americans also ignited fires. But whereas Indigenous people had a schedule ("cool burning" in spring and fall), Europeans recklessly burned slash piles – debris from the cutting of trees – even when it was hot, dry, and windy. The burning of slash piled up by farmers and lumber companies in New Brunswick was likely behind the Miramichi Fire. But newspapers also blamed comets, spontaneous combustion, and the rays of the sun. According to the Montreal *Gazette*, it was "perfectly evident that this fire, like that which lately ravaged Upper and Lower Canada, must have been occasioned by the rays of the Sun, operating on some inflammable matter on the surface of the Earth during an unparalleled hot and dry summer."[8]

3.2 Fires have become common since entrepreneurs began draining the Great Dismal Swamp from the 1750s to the 1960s.

Historian Alan MacEachern must have had some fun going through digital files to find newspaper coverage of the fire. In his very fine book on the Miramichi fire of 1825, he noted how the *Gazette* mocked its rival for suggesting that the forest had burned spontaneously. In response, the *Gazette* opined that "a courier from the planet Mercury, which called in as the principal towns of Venus and the Moon, had informed it that these locales had been on fire that summer too, which we attribute to their combustive matter."[9] A writer for the *London Examiner* took the situation a little more seriously, blaming the fires on the "carelessness" of the Mi'gmaq. It's possible they were to blame, but no one really knows. New Brunswick historian Peter Fisher was so sure that "Indians and lumberers" who "roam through the wilderness" were responsible that he expressed surprise in an 1838 article that fires as big as Miramichi did not happen more often.[10]

They hadn't up until then. Studies suggest that large fires were largely absent from the forests of northern Maine until that time as well. No one had ever seen a complex of fires so massive it burned across 15,500 square kilometres of forest. The exact size will never be known because "the suddenness of the calamity at Miramichi," as the anonymous writer of a charity pamphlet observed at the time, "prevented a cool and collected observation of facts."[11] What is known from this account and others is that the summer and fall of 1825 had been extraordinarily warm. No rain of significance had fallen between early July and 8 October. Dairy and beef farmers managed to get by, as did other farmers who grew fruits and vegetables. Yields, however, were low, and harvesting was made difficult in many cases because of an outbreak of yellow fever and typhoid – a waterborne disease thrives when water levels are low in streams and rivers.

What may have made the situation worse, according to some, was the spruce budworm that periodically invaded the east coast forests. If that were the case, then the stands of dead trees, along with slash piles in a tinder-dry forest baking in hot weather may have provided more than enough fuel for small fires to spread rapidly in and around Miramichi and in the state of Maine to the south. Several small towns perched alongside the shores of the Miramichi River found themselves at the epicenter. The largest were Chatham, Douglastown, and Newcastle, which was later renamed Miramichi.

The premonition of a catastrophe, however, didn't arrive until the afternoon of 7 October, when a column of smoke was seen rising in the near distance, just northwest of Newcastle. No one was too concerned initially because they had never experienced a fire that had caused them harm. Indifference turned to alarm at 7 p.m. that night when, according to the author of the charity report, "a smart wind" blew in with "ashes and cinders in such quantities that those persons who were exposed were nearly blinded and suffocated before they could retreat under cover. The inhabitants kept within doors, and many retired to their beds as usual with them on dark nights."[12]

The author also noted that "They had no fears of any further consequences than the temporary inconvenience,"[13] but by 8 p.m., almost everyone realized they were in serious danger. As the winds reached hurricane force, a roaring noise got louder and louder, to the point where some villagers thought that "the earth had loosened from her ancient foundations." No one knew what to do, except to gather buckets of water.

"The people seemed to perfectly stupefied," recalled Robert Cooney, who lived within a mile of Newcastle. "Every body seemed to be alive to the danger;

but no one seemed capable of warding it off."[14] The screams of the burnt and the wounded mingled with the cries of domestic animals scorched and suffocating in the heat."[15] Men emerged from their homes half naked. Women followed with children wrapped up in blankets. Those stricken with typhoid or yellow fever were left to find safety. There were no fire brigades to come to the rescue and no railway companies to transport people out of harm's way. The only two places that offered refuge were a nearby marsh and the river. Most people chose the river. Some hopped into boats; others waded up to their necks in the water, trying to coax their cows and horses to come in with them.

The livestock were so panicked by the smoke and the flames that they refused to enter. Wild animals, however, had no such fear. Soon humans in the river found themselves surrounded by wildlife, including racoons, deer, and even large moose. In one case, a bear was paddling among them. The bear refused to budge until the fire passed.[16]

Even the fish were overwhelmed by the ash and debris that poured into the river as the fire raged. Salmon runs, which were already depleted by foresters' overfishing – were almost non-existent for the next two years.[17] Several ships that were anchored in Miramichi Bay caught fire when a dazzling shower of embers ignited lumber piled on their decks. Unable to hoist sails because of the hurricane-force winds, several of the burning ships were set adrift, bringing fire to the opposite side of the river. The master of one of the ships took up a bible, firmly believing that the "Last Day" had come. "Men women and children were heard screaming in every direction, some rushing from the fire bemoaning the loss of their husbands, children, parents, wives, friends, and many suffering under the most excruciating agonies from the burns which they received."[18]

Local stories told later by people who survived the fire of knew something of it spoke to both the horror of the event and to the cultural mores of the time. A "coloured" girl who had been imprisoned for running away with an illegitimate child, presumably her own, was released from jail to fend for herself. She did not survive. All nine members of one family perished. Only two of another similarly sized household survived, but were forced to bear witness to bodies with heads burned off, brains exposed to view, and bowels bursting. One man left his three children at home so that he could plunder the homes of his unfortunate neighbours. His children died. A more honest man carried his sick wife and children to safety. He refused the offer of food until his wife and children and others were served.

"A greater calamity, the Fire … never befell any forest country," Cooney noted in 1832 his first-hand account of the fire. "The general character of the scene was such, that all it required, to complete a picture of the General Judgment, was the blast of a TRUMPET, the voice of the ARCHANGEL, and the resurrection of the DEAD."[19]

Fire did not stop at the Canadian border that year. Approximately 3,400 square kilometres of Maine also burned. Thanks to the sleuthing of Alan MacEachern, we know now with some certainty that many of the fires in that state and those in New Brunswick joined to produce one of the largest conflagrations in North American history.[20] Apparently, no one died in Maine, but at least 165 people lost their lives in New Brunswick. Most were laid to rest where their charred bodies were found. About one-fifth of the future province of Canada was burned. Many of those who survived were disabled and left destitute because, at that time, money was stored at home rather than in banks.

What followed was one of the bigger humanitarian relief efforts in early-nineteenth-century North American history. Churches gathered collections at Sunday services. Citizens, both rich and poor, chipped in with donations. Soldiers gave up a day's pay. His Majesty's Council, in conjunction with the government assembly, poured money into a relief fund. A war ship was deployed to bring food and supplies to the needy. As with modern-day disaster responses, Canadian decision-makers were happy to give money for the relief efforts. What they were not prepared to do was come up with a plan to deal with future fires, as Americans did with fire-control laws that were put in place in some parts of the northeast by the time of the Revolution.

This lack of response wasn't because people didn't want to do anything. The problem was that they didn't understand the physics of fire and the biology of succession because the science was still in its infancy, and much debated. In addition, there was a deep divide between science and religion. Great calamities were often seen as evidence of divine judgment, not human error or natural causes.

If the Dark Day of 1780 was a warning about the inherent dangers of fire, the Miramachi Fire signalled the coming of a new era of wildfire in North America where human settlement and developments intermingled with forests and grasslands with little regard for defensible spaces, or for the fact that wind can blow embers great distances. Lawmakers and landowners continued to be indifferent or oblivious to the impact of trapping beavers and draining bogs, fens, and marshes in a warming climate. Railway companies, like everyone else, ignored the danger of clearing the land and were oblivious to the possibility of another catastrophe.

On her travels through the backwoods of Canada in the 1840s, Catharine Parr Traill, Susanna Moodie's sister, noted that "man appears to contend with the trees of the forest as though they were his most obnoxious enemies, for he spares neither the young sapling in its greenness nor the ancient trunk in its lofty pride; he wages war against the forest with fire and steel."[21] She was contemptuous of those who attributed the hazy orange-sky "Indian summer" days to First Nations overzealously burning the forest. Those smoky skies that blotted out the sun, she said, were the work of settlers burning brush. Fires displayed their brute force many times before Parr Traill died in 1899. In 1845, vast fires west of Lake Superior to Rainy Lake in Ontario burned an estimated 269,000 hectares. A decade later, a conflagration twice that size torched the Temagami region.

The Ottawa Valley of Ontario didn't burn nearly as big in August 1870, but it got the most attention back then even though it is almost forgotten today. Little rain had fallen that summer and temperatures were, as the *Perth Courier* put it, 85 to 90 degrees Fahrenheit in the shade. The trouble started eighty-five kilometres southwest of Ottawa in the Perth area along the Canada Central Railway. Extreme as the fire danger was at the time, railroad workers had decided to burn ties, fencing, and other construction materials. When a wind came up and sent the embers into the cedar, spruce, and balsam forest, they tried in vain to put it out. By the time railroad cars arrived with barrels of water, it was too late to do anything. Gusty winds sent the fire racing out of control towards Ottawa, where citizens were oblivious to what was going on until 2,000 homeless people began marching in.

"It is utterly impossible to conceive of the misery and desolation that exists in many places, both in the Counties of Lanark and Carleton," a reporter wrote in the *Perth Courier* newspaper. "So rapid and overwhelming did the devouring element often become, that the terrified people were glad to escape with their very lives leaving everything behind ... even to a scanty supply of wearing apparel ... to the rapacity of the dreaded monster. Horses, cattle, sheep, pigs, and even dogs, have, in some instances, become easy victims to the fire, escape having been rendered impossible from the suddenness of its appearance and the utter inability of the weak efforts of man to subdue it or keep it in check. 'The roar of the fire,' said one who heard it 'was terrible.'"[22] An Irish settler who was identified only as Mrs Patrick Hardin was found on the bank of a creek holding a world clock that had presumably been passed on from one generation to another. Another farmer died after he returned to his home to retrieve some papers. The fire burned so intensely in some places that evidence of the fire can still be seen in Burnt Lands Provincial Park thirty kilometres west of Ottawa.

On 19 August, the *Ottawa Times Weekly* reported that some 2,000 homeless and hungry people were on Richmond Road making their way to the capital. A headline in *The Globe* the next day announced, "Panic in Ottawa." Buildings in the Lansdowne Park area of Ottawa burned down before someone in the Ottawa Fire Department got the idea of opening the St Louis Dam at Dow's Lake to flood the streets. As the fire raged out of the city into rural villages and farms, no one knew what to do. Like the citizens of Miramichi, they had never had to deal with an inferno like this. Many jumped into the Mississippi or Ottawa rivers. Others climbed down ladders into their wells.

The Globe reported that "it was difficult to convey to those who have not seen a bush fire, the terror, the anxiety, and the mad excitement it occasions. It raises itself a tempestuous wind, which hurls burning leaves, loose sticks and branches of trees to considerable distances; and these alighting on roofs of houses and brans, thoroughly dried by wind and sun, soon ignite buildings."[23] The correspondent insisted that "it is full time that something be done to prevent the recurrence of these periodical fires. The Provincial Legislature should take the matter up and frame measures to restrain settlers from setting fire to the woods, even for such a laudable purpose as that of clearing the land."[24]

Government officials did not respond to the disaster as generously as authorities had in New Brunswick in 1825. Sir Francis Hincks, the federal minister of finance, refused to offer aid, insisting that it was a provincial responsibility. For their part, lawmakers in Ontario dismissed a proposal for financial relief as a waste of money. J. Sandfield Macdonald, the premier of the province, went so far as to suggest that the afflicted must have brought the disaster on themselves for God Almighty to have acted this way. A provision of aid, he added as an excuse and ill-informed afterthought, would be unconstitutional. When public opinion turned angrily against him, he offered a 6 per cent loan for those who could offer security. Most, of course, didn't qualify, because they had nothing left of value. Those who did could get a better deal at many banks.

Macdonald was eventually shamed into being more generous when he was reminded that the Quebec government had offered to compensate each resident of Saguenay who had lost everything to fire that same month. The fire in the Saguenay-Lac-Saint Jean region had also been started by farmers clearing the land. Seven people died there and twenty lost their lives in Ontario. Many more were seriously injured. The damage done was incalculable.

The 1870 fires were big news in their day. News of the Saguenay and Ottawa Valley fires made it on to the pages of the *San Francisco Chronicle*, *The Globe*, and the *Toronto Daily Star*. "All of Canada in a Wall of Flame," the *Perth Courier* headline read. The editor was not far off the mark. *The Globe* noted that summer fires were burning simultaneously in Thunder Bay, Saguenay, New Brunswick, Prince Edward Island, and throughout the state of New York. Sufficiently impressed with the forceful nature of the fire around Thunder Bay, *The Globe* noted how the flames and embers

> sprang up trees and blasted away their leafy greenness, left behind in a moment smoking and blackened trunks. It swept across the swamp that was vainly regarded as protecting the clearing, lit up for a moment the humble homestead of a settler and ruthlessly blotted out in the wink of an eye the fruits of labour. It sprang across broad rivers as if they were rivulets, consumed bridges and earth covered culverts. The cordwood stacked up in the beach, the shanty of the half breed, the dwellings of the miners, all were swallowed up in an instant, women and children having barely

> enough time to rush from their homes and take shelter in the mouth of the silver mine, while all their fowls, dogs and worldly property were swept away. Such was the great forest of Thunder Bay.[25]

3.3 The Saguenay Fire, image from the *Canadian Illustrated News*, 25 June 1870. The one-day fire killed seven people and left hundreds of families homeless.

No such headlines were put to immediate print a year later when the deadliest wildfire in North American history swept through northeastern Wisconsin, burning 485,000 hectares. There were several reasons why the Peshtigo Fire was overlooked and largely unreported for a few days. Much of Wisconsin on both sides of Green Bay was aflame. Most of the telegraph lines that would have sent word of the blaze had burned down. Even if the lines had been intact, the media's resources were focused on a fire in Chicago that levelled roughly nine square kilometres of the city, killing about 300 people and leaving 100,000 homeless that same night. That fire had nothing to do with what was happening in the forests. Legend suggests that it started when a cow knocked over a lighted lantern. But Patrick and Catherine O'Leary, the cow's owners, denied it, and the true cause has never been determined.

Chicago burned big and quickly because of an abundance of wooden buildings that made the city just as vulnerable as the wooded areas of other parts of Illinois and Michigan where the Port Huron Fire burned through the area known as the "Thumb" of the state. At least fifty people died there. "A sky of flame, of smoke a heavenful, the earth a mass of burning coals, the mighty trees, all works of man between and living things trembling as a child before a demon in the gale," is how the Port Huron Fire was described in a history of Sanilac County. "To those who have seen, the picture needs no painting."[26]

Peshtigo, located on the banks of the river of the same name, was home to country's largest woodenware factory and several sawmills. Residents took pride in the fact that it had a foundry, stores, hotels, a schoolhouse, and two churches: one Protestant, the other Catholic. Like many Midwest towns, it had been built on exploitation of the surrounding forest.

No one was overly concerned on the morning of 8 October when they woke to the smell of smoke and a red glow in the sky. It was an all too familiar sight at that time of year. But as Father Peter Pernin, the Catholic priest, later noted, attitudes changed when the smoke got thicker, the skies got darker, and when a "distant roaring" broke the "preternatural silence reigning around."[27] The wall of smoke and flame came in so quickly that Pernin and almost every one of the 1,700 people in Peshtigo found themselves in the same situation as those in Miramichi had a half-century earlier. They didn't know what to do or where to go because they had never experienced anything like this before. Rapidly rotating columns of hot air containing smoke, flames, and fire debris whirled in like tornados. People who try to flee across the bridge to the other side of the river found that the forests there were also burning. Many jumped into the river when vehicles, cows, and pigs on the bridge blocked their way and added to their confusion.

Pernin himself was nearly blinded by a blast of hot air when he entered the burning Church in order to save the Holy Eucharist. He made it out with the Holy Eucharist in hand, but just barely, and to little relief.

> The air was no longer fit to breathe, full as it was of sand, dust, ashes, cinders, sparks, smoke, and fire. It was almost impossible to keep one's eyes unclosed, to distinguish the road, or to recognize people, though the way was crowded with pedestrians, as well as vehicles crossing and crashing against each other in the general flight. Some were hastening towards the river, others from it, whilst all were struggling alike in the grasp of the hurricane. A thousand discordant deafening noises rose on the air together. The neighing of horses, falling of chimneys, crashing of uprooted trees, roaring and whistling of the wind, crackling of fire as it ran with lightning-like rapidity from house to house – all sounds were there save that of the human voice. People seemed stricken dumb by terror.[28]

Peshtigo was one of seventeen villages in Wisconsin levelled by fire that fall. What set the fire there apart from others was that it killed between 1,200 and

3.4 *The 1871 Peshtigo Fire*, oil on fibreboard by Mel Kishner, 1968.

1,500 people: the most in North American fire history. "Many circumstances tended to prove that the intensity of the heat produced by the fire was in some places extreme, nay unheard of," Pernin wrote, continuing:

> the flames pursued the roots of the trees into the very depths of the earth, consuming them to the last inch. I plunged my cane down into these cavities, and convinced myself that nothing had stayed the course of combustion save the utter want of anything to feed on. Hogsheads of nails were found entirely melted though lying outside the direct path of the flames. Immense numbers of fish of all sizes died, and the morning after the storm the river was covered with them. It would be impossible to decide what was the cause of their death. It may have been owing to the intensity of the heat, the want of air necessary to respiration – the air being violently sucked in by the current tending upwards to that fierce focus of flame – or they may have been killed by some poisonous gas.[29]

Peshtigo was for some time a mere footnote in the history of the Midwest, partly because it was overshadowed by the Great Fire that burned most of the city of Chicago almost simultaneously, and because the forests of so many other Great Lakes states burned that fall. As more conflagrations occurred in the years that followed, the significance of the Peshtigo Fire was lost in an historical record in which politics and personality rather than nature dominated. Novelist William F. Steuber turned it into a novel that was favourably but briefly reviewed by the *Chicago Tribune* in 1957.[30] Not until 1968, when journalist

Robert W. Wells wrote on the subject for the *Milwaukee Journal*, was there a skilful and readable reconstruction of what had happened. Pernin's full account was published three years later, not in a national journal such as the *Smithsonian Magazine* but in the *Wisconsin Magazine of History*. Ironically, what was once known as the "Forgotten Fire" has become one of the most famous and well-studied, thanks to historians such as Stephen J. Pyne.

Miramichi was not so quickly forgotten, as MacEachern notes in describing how new stories – "maudlin, fantastic and symbolic" – emerged in the decades that followed. One story tells of thirteen babies being born on rafts on the night of the fire. Another tells of a young woman who ended up marrying the man who resorted to "superhuman efforts" to save her when she fell off a raft. The best is about a cow that carried a two-year-old girl across the Miramichi River to safety.[31] The second best is of an insurance company that began offering fire insurance in newspaper advertisements a week after the fire. The ad, as MacEachern notes in his book, ran for years.

George Perkins Marsh refers to the Miramachi in *Man and Nature*, his seminal 1864 treatise on the impacts that humans have on the natural world. The subject of Miramichi was still appearing on exams at Dalhousie University in 1885. But the memory faded until 1959, when American folk singer Edward "Sandy" Ives wrote a song about it; subsequently novelists Annie Proulx in *Barkskins* (2016) and Valerie Sherarrd in *Three Million Acres of Flame* (2009) brought it back to life in fiction. Fire historians did not forget, as they seem to have done with the Ottawa Valley fires and several others notable for the impact they had on people. But the fact that it took 195 years to get MacEachern's full accounting of the Miramichi fire in a history book says a lot about our strange and unsettling relationship with fire.

4

THE BIG BURN

In the four decades that followed the Ottawa Valley, Peshtigo, and Midwest wildfires, it was becoming obvious to anyone who was paying attention that a new paradigm of fire was playing out in North America. One of those paying attention was John Wesley Powell, a geologist, soldier, and ethnologist who would lead the first US government-sponsored expedition through the Colorado River into the Grand Canyon before becoming the second director of the US Geological Survey. In his 1878 *Report on the Lands of the Arid Region of the United States*, Powell offered up this bit of advice: If fire could be prevented, the annual growth of trees would more than replace the number cut down by the timber industry.[1]

This was a tantalizing prospect for those in the timber industry who recognized that there might be an end to the forests they were harvesting. Clearcutting had reached its peak in New England, Virginia, and North Carolina by 1880, when most of the white pine forests were gone. By then 60 to 80 per cent of New England had been cleared for pasture, tillage, orchards, and townsites.[2] Timber companies, which were forced to move to the Midwest shortly afterwards, were soon on the march to the Pacific. There was little effort to reseed when it was easier to move on to the next virgin forest.

Franklin B. Hough, a New York physician who had a longstanding interest in the natural sciences, echoed Powell's sentiments in 1882. After conducting the first national investigation of wildland fire on behalf of what would become the US Forest Service, he declared that "there is no subject in Forestry more important than the prevention and control of forest fires." It wasn't the nature of timber harvesting that concerned him. It was fire. Fire, he asserted, caused "greater destruction to the forests than that occasioned by all the demands of industry put together."[3]

Seeing fire as the enemy, rather than as a natural process that leads to death, birth, and renewal, resonated with others as high-intensity fires increased, taking their toll not only on valuable timber but on people in harm's way.

The list of consequential fires was becoming a long one. Fire burned the toe of Michigan again in 1881, destroying 1,480 barns, 1,521 houses, and 51 schools, while killing 300 people and injuring many others. Newspaper reports described how towns like Bad Axe, Verne, Forest Bay, Richmondville, Charleston, Anderson, Deckerville, Harrisville, and Sandusky were completely burned down. Women and children were found in fields nearly naked.[4] News of this fire, however, was overshadowed by others in New York, Pennsylvania, and Ohio, where 15,000 men were fighting conflagrations day and night the same week. Those in Pennsylvania became especially spectacular when they overran forty-five oil rigs. More than 100,000 barrels of oil were consumed by flames.

A week later, fire burned out of control in the Parry Sound and Muskoka districts of Ontario. Newspapers reported that it had consumed 518 square kilometres of forest and a million metres of saw logs owned by the Cool Bros lumber company. There were many more fires that year, but government officials couldn't tally the full extent of the conflagrations because they did not take into account grass and bush fires.[5]

Another that stood out was a blaze in early September that produced a sky over Toronto that was of such "strange, weird glory" that the *New York Times* reprinted an article from *The Globe* describing people's reactions. Like many of those who had experienced the Dark Day at Noon a little over a century earlier, many Torontonians interpreted this as a sign that "the last trump was about to sound." Those who took refuge in their homes that afternoon were reportedly "weeping and wailing" by candlelight. The next day Bostonians woke up to the same sight, presumably caused by the same fire.[6]

One of the great legacies of the 1881 fire season was Clara Barton founding the American Red Cross. The sixty-year-old set it up to provide money, food, clothing, and furniture to some 14,000 people in Michigan. The relief effort was so effective that President Chester Arthur and the US Senate officially recognized the Red Cross in the signing of the Geneva Treaty the following year.

The Hinckley Fire of 1894 in northern Minnesota was notable for killing 418 people, not including native Americans who also died and were not counted in the official record. The fire burned between 80,000 and 100,000 hectares in less than four hours. More than a quarter of the victims died in a swamp that offered no refuge. Many more would have perished had it not been for a train arriving just in time to transport people away from the area. All that was left were the walls of a schoolhouse, the iron fence around the Town Hall property, the bank vault and one building that somehow survived. The story was covered in the local St Paul paper ("Inferno in Forest"), in the *New York Times* ("Hundreds Perish in Fires"), and overseas in British papers ("Great Forest Fires in America").

In 1897, fire devastated two townships on the North Dakota side of the Turtle Mountains along the Manitoba border. Fires on the Manitoba side spread so far and quickly that men and beast were paid no attention to each other as they fled the flames.

In 1902, the Yacolt Fire burned in what is now the Gifford Pinchot National Forest in Washington State, killing thirty-eight people. Flames were reported to be 100 metres high and moving at rates of sixty kilometres per hour. One headline screamed out the news for the rest of the world to hear. "Fatal Forest Fires Rage ... People Take to the Rivers to Save Their Lives."[7] Then, in 1908, 95 per cent of the British Columba town of Fernie, just north of the Montana border, was levelled. Twenty-two people died and nearly 10,500 hectares of forest was burned. It was the second time in five years that the town was damaged by fire.

One of the unnerving things about all these fires was the speed with which they moved. The Yacolt Fire was not the only one that behaved with such fury and velocity.[8] In the fall of 1901, a fire near Queenstown in southern Alberta "travelled at a rate of forty or sixty miles per hour."[9] Another in central Nebraska travelled 200 kilometres in a single day. An 1887 fire in Texas travelled more than twenty-seven kilometres in two hours. Farmers and ranchers could do little to stop these fires except to make it a cause for dismissal if an employee failed to immediately help put them out.[10] Most, of course, had didn't have enough employees to deploy.

Forest professionals and some in the railroad and lumber industries recognized that "business as usual" was not going to be successful in preventing future catastrophes. Some in the business community even dared to ask the question that John Wesley Powell posed in 1878, "Can these forests be saved from fire?"[11] In 1910, California timber owner George Hoxie articulated many of these concerns by suggesting that light burning – the Indigenous art of managing forested landscapes – should be considered a tool to control fire, as it already was by people in the south where the practice was called "woods burning." "We had best adopt fire as our servant; otherwise it will be our master," Hoxie wrote in an article in *Sunset Magazine*.[12] The state engineer of California was in favour of it. A lot of timber owners warmed to the idea. Officials with the Southern Pacific Railroad company were on board. Even Secretary of the Interior Richard Ballinger was supportive, stating that "we may find it necessary to revert to the old Indian method of burning over the forests annually at seasonable periods."[13]

Hoxie and Ballinger might have prevailed had a fourth element – politics – not been added to the Fire Triangle in 1910. Politics had first come to the fore

4.1 Gifford Pinchot, fourth chief of the Division of Forestry, 1898–1905, and first chief of the Forest Service, 1905–10.

in 1905, when responsibility for the Forest Service was transferred from the Department of the Interior to Agriculture. With that move, the agency was given responsibility for an area of land larger than most European countries. Mining and timber barons weren't happy. US Senator Weldon Heyburn, a lawyer representing a mining interest, made it his mission to stop President Teddy Roosevelt from further expanding the reach of the Forest Service into the west. If Heyburn had had his way, the Forest Service would have been abolished, but he was outwitted by his nemesis, Gifford Pinchot, who headed up the Division of Forestry for seven years before it morphed into the US Forest Service in 1905. A Republican at heart, Pinchot was as politically connected as anyone in government at the time, defined by bloodline and a monied family that had made its fortune in lumbering, land speculation, and wallpaper.

Unlike his German-born predecessor Bernhard Fernow, who had no forests to manage and who considered himself to be more of an educator and researcher than a political operator, Pinchot was often outspoken about what needed to be done and sometimes brutal working behind the scenes to get what he wanted. Like Franklin B. Hough, who Pinchot described as "perhaps the chief pioneer in forestry in the United States," he equated forest health to the moral health of the nation. And like his friend Madison Grant, a fellow Yale Law School graduate, he believed that management of forest and wildlife was a measure of character.

Pinchot often got his way because President Theodore Roosevelt, a close friend – they boxed (Pinchot won) and skinny-dipped together in the Potomac – regarded him as one of his most trusted public servants. Pinchot, according to those who knew him, was Roosevelt's conscience on conservation. In fact, Pinchot wasn't a "conservationist" in the modern sense of the word. Big animals such as bear and elk mattered to him, but insects and small animals did not. He saw forests as a utility which would provide a reliable source of income for the country. Like John Muir, another friend and the most famous naturalist of the time, he had little regard for the way lumbermen like his grandfather had exploited that resource. (Pinchot's father sorely regretted what the family business had done to the forests of Connecticut, and it was that lament that shaped this view.)

"When the Gay Nineties began, the common word for our forests was "inexhaustible," he once said. "To waste timber was a virtue and not a crime. There would always be plenty of timber ... The lumbermen ... regarded forest devastation as normal and second growth as a delusion of fools ... And as for sustained yield, no such idea had ever entered their heads. The few friends the forest had were spoken of, when they were spoken of at all, as impractical theorists, fanatics, or 'denudatics,' more or less touched in the head. What talk there was about forest protection was no more to the average American that the buzzing of a mosquito, and just about as irritating."[14]

Pinchot did not see light burning as the solution to the wildfire problem that threatened these forests. Like Muir, who lamented the "strangely dirty and irregular life" of Indians,[15] he saw nothing in their management of the forest and prairie as worth copying. It was the lack of a strategy and a shortage of fire suppression resources that was problematic. "The one secret to fighting fires is to discover your fire as soon as possible and fight as hard as you can and refuse to leave it until the last ember is dead," he told the *New York Times*.[16]

Pinchot also made short work of Interior Secretary Ballinger's ideas, not just by attacking the strategy of light burning that Ballinger and Hoxie promoted, as he did so effectively in newspapers such as the *New York Times*, but by publicly denouncing him for virtually giving away federal coal reserves. Pinchot agreed with and possibly informed a *Colliers* magazine article that accused Ballinger of improperly using his office to help the Guggenheims and other powerful interests illegally gain access to Alaskan coal fields. Ballinger was not guilty in a strictly legal sense; the situation was a little more complicated than it appeared to be. But due to the public outcry, President William Howard Taft was left with no choice but to fire Ballinger in January 1910. Unwilling to forgive Pinchot for

4.2 *Ploughing the Fields That Flames May Be Starved*, drawing by Cyrus Cuneo from *The Illustrated London News*, 22 October 1910.

his insubordination, he fired him as well, driving a wedge between Republicans who supported him and those who supported Roosevelt.

One of the events that put an end to the idea of "light burning" was a prescribed fire in California that got out of control in the summer of 1910, blackening more than 13,000 hectares. It could not have happened at a worse time. In May, newspapers were reporting fires in Minnesota, Michigan, Wisconsin, Alberta, and Manitoba.

The other more important one was a huge complex of fires that ripped through the northern Rockies in the United States and southernmost parts of British Columbia and Alberta just months after Pinchot and Ballinger were dispatched from their jobs. "The Big Burn," "The Great Burn," or "The Big Blowup," as that 1910 event came to be called, was famous not only for the death, damage, and destruction that came with it, but for the pall of smoke it sent to the east coast, darkening cities such as Boston.[17]

The Big Burn ushered in an era of wildfire suppression that would leave an indelible mark on the North American landscape for more than a century. According to Stephen J. Pyne, virtually every story of fire protection on public lands can trace its origins to what happened in 1910. Firefighting would no longer be a novelty that most rural Americans distrusted, but a trusted strategy that was no different than the art of fighting fires in cities.[18] Never mind that it would eventually fall short, as some in the US Forest Service predicted. Suppressing fires became the equivalent of war, where victory was perceived to be assured so long as there was money, men, transportation corridors and technology, and ranger stations situated a day or two horse rides from each other. It wasn't just Pinchot who believed it. Newspapers such as the *New York Times* also came to the conclusion that most fires were "ordinary fires, which if caught early, can be promptly suppressed before they attain serious proportions."[19]

The conflagration began with a complex of more than 1,700 small fires ignited by lightning, loggers, homesteaders, and campers in the Bitterroot Mountains along the Idaho-Montana border. The newly constructed Chicago, Milwaukee and Puget Sound Railway, which had a line running through the region, was one of the culprits. One report estimated that coal-powered locomotives shooting out red-hot cinders triggered more than one hundred of these fires into a tinder-dry forest that had not seen rain all summer. Literally, all hell broke loose when hurricane-force winds sent flames and embers shooting thirty metres into the sky and in all directions; "Truly a veritable demon from hell," is how one forester described it.

An estimated 1.3 million hectares of trees were torched in just two days. Eighty-six people died and several small towns were destroyed. Among them were Grand Forks, Idaho, and Taft, Montana, where a fifth of the population was made up of prostitutes working in saloons. In the towns of Avery and Wallace, which were also destroyed, apples baked on trees, and window glass on buildings melted. Its residents would have suffered a similar fate had a train and the black Twenty-Fifth Infantry, also known as the Buffalo Soldiers, not come in to evacuate them.

There was a small army of forest firefighters that were enlisted two weeks before the fire began to make its big run. This was due in part to the 1908 passage of the Weeks Act that laid the foundation for the establishment of national forests in the eastern United States and authorized the Forest Service to spend whatever was necessary to combat blazes. Congressman John Weeks, ironically, once declared there would "not be a cent for scenery" coming from Congress.

Firefighters even had an instruction manual written the year before by Henry Graves, who was second in command of the Forest Service before he took over for Pinchot. But all the brigades amounted to were jacks-of-all trades who rode horses, if they had them, and were poorly paid, subjected to long periods of isolation, and faced with seasonal layoffs. Money for the newly enlisted was so long in coming that rangers like Ellers Koch used their own savings to pay some of the new recruits. But because there were not nearly enough of them, jails in

4.3 Men burned in the Big Burn of 1910.

Idaho were emptied,[20] and crew bosses like Ed Pulaski were forced to go into saloons and brothels to recruit whoever was able to fight the fires. According to the *New York Times*, women were relied on to extinguish fires at home as they did heroically when fire came to Elk City in Idaho.[21]

The best tool firefighters had was to "backfire" – a strategy that made untrained men nervous because they couldn't see how lighting a new fire in the path of an existing one would stop the momentum or change the direction of a conflagration, as it does when properly executed. But even with backfiring, the military, like the 9,000 to 10,000 men stationed on the front lines of the Big Burn, didn't stand a chance of slowing the firestorms that came at them with ridiculous speed and suffocating smoke. According to Ferdinand (Gus) Silcox,

4.4 Coeur d'Alene hardware warehouse ruins after the Wallace fire of 20 August 1910.

the assistant district forester of the region, "the roar from these fires was heard for miles and was likened by some of the Rangers in their path to the noise of a thousand freight trains crossing simultaneously as many steel trestles. At many points these fires jumped rivers a quarter or half mile wide, and in several instances leaped across canyons a mile or more in width, from ridge to ridge, leaving solid strips of green timber untouched."[22]

Seventy-eight firefighters were killed. One crew of forty-three firefighters, dispatched to the wilderness outside of the town of Wallace, would have also succumbed had they not been forced to flee under Pulaski's orders. They barely made it to the shelter of a mineshaft. Exhorting the men to lie down on the tunnel floor while he hung wet blankets over the entranceway, Pulaski threatened to shoot anyone who tried to make a run for it. All but five of the men survived.

Smoke from the fires darkened the twenty-first day of August as far north as Saskatoon, as far south as Denver, Colorado and as far east as Watertown, New York.[23] The fire was such a horror story that War Department was asked at one point to bombard the skies to induce clouds and allow rain to fall. It declined, suggesting the estimated $100,000 price tag would be a waste of money.[24] That said, soldiers who were sent to the town of Wallace to help fight the fires tried, aiming dynamite and cannon fire into the sky for sixty straight hours.

When naturally occurring rain finally extinguished the flames, Pinchot, still smarting from his dismissal, issued a formal statement blaming congressmen such as Weyburn for doing everything in their power to undercut the Forest Service. "If even a small fraction of the loss from the present fires had been expended in additional patrol and preventative equipment, some or perhaps all of the loss could have been avoided," he stated. "Forest fires are preventable. It is a good thing for us to remember at this time that nearly or quite all of the loss, suffering and death these fires have caused is wholly unnecessary."[25] A fire in the forest, added Pinchot, is like a fire in the city. A fire department is needed in both places to extinguish the fires while they are "still young."

In assessing the damage done by the Big Burn, Henry Graves, who was appointed chief of the US Forest Service in 1910, echoed Pinchot's call for more an aggressive approach to fighting fires.[26] This was no surprise. They were old friends. Like Pinchot, Graves came by way of Yale University, where he was the first professor and director of the Yale Forest School. It was Pinchot who had recruited him to the Forest Service and made him second in command. The two did not differ in their views on fire; they just went about it differently. Where Pinchot was happy to be front and center in public debates, Graves liked to work behind the scenes.

There was no doubt about where Graves stood on the matter of light burning. Forest fire protection is "the first measure for the successful practice of forestry," he proclaimed. "The doctrine of light burning ... is nothing less than the advocacy of forest destruction, and those who preach the doctrine have a large share of responsibility for fires which their influence has caused."[27]

5

BIG BURNS IN CANADA

Within a year after the Big Burn played itself out in the newspapers, Congress defeated a bill that would have privatized much of the burned forest, and passed the Weeks Act, which authorized the government to purchase up to thirty million hectares of land in the east to protect watersheds from development and wildfire. The act mandated the US Forest Service to work with state fire bureaus, which was happy to cooperate because co-operation came with funding it could not otherwise afford. Any notion of using prescribed burning to manage the forest was doomed. From that point on, fires would be fought with military precision on a near-national level.

The Canadian government followed suit that same year with a new forest policy that added three more forest reserves to the twenty-one that were already in place, as well as hiring more firefighters and forest rangers. Canada had already experienced a number of large destructive fires following the one in Fernie in 1908. In 1910, the Baudette–Rainy River (aka Spooner-Baudette) fires burned 121,500 hectares, including a town of 1,500 people along the Ontario-Minnesota border. An even bigger blaze in southwestern Alberta that year went virtually unnoticed by the rest of the country because only a small number of ranchers and homesteaders were in harm's way.[1] So much of Canada was burning in 1910 that *The Globe* declared wildfire to be a "national menace."[2]

As deadly as the Baudette–Rainy River fires were, with forty-two people killed and more than a thousand left homeless, the Porcupine Fire of 1911 was in a class of its own. It burned with such speed and intensity that it drove six

5.1 The Fernie Fire of 1908.

hundred people into a lake to seek refuge from flying embers and noxious smoke. News of fires in the Porcupine Lake region began on 11 July 1911, when telegram reports clicked-clacked across the wires of the country with stories that described "horror after horror." "NORTHERN ONTARIO SWEPT BY WALL OF FIRE" was the headline in *The Globe* that day. "The Towns of Cochrane, South Porcupine and Pottsville Have Been Wiped Off the Map – Flames Extend for Hundreds of Miles."

Nothing like this had happened in Canada since the Miramichi Fire of 1825. Those who chose not to seek refuge in the lake didn't stand a chance when they found themselves in the path of that wall of flames. In one case, a fire whirl or possibly a fire tornado (pyrogenetic tornado) sucked the oxygen out of the atmosphere with such force that mature trees were ripped from the earth. Whirls form when rising hot air twists flames into a towering pillar that can, in some cases, lift heavy objects off the ground. Many of those who thought they would be safe in a mine shaft ended up suffocating. Others drowned when a boxcar full of dynamite exploded by the shores of Porcupine Lake. The wealthy owner of one of the mines in the region died while trying to save his cat.

The fire couldn't have come at a worse time for the Ontario government. Just a month earlier, it had placed several articles in *The Globe* inviting people to move into the region to take advantage of the six to nine million hectares of arable land that had opened up with the recent completion of the Temiskaming and Northern Ontario Railway. In the articles, director of colonization and immigration Donald Sutherland pointed out that "the chance of crop failure through lack of sufficient rainfall is very small." He also suggested that the "prodigal manner in which vast areas were allowed to be devastated by the lumberman and by fire" was a thing of the past.[3]

5.2 The Porcupine Fire killed at least seventy people on the north side of Porcupine Lake near Timmins, Ontario, in 1911. Photo from the *Toronto Daily Star*.

In truth, Sutherland, and pretty much everyone in government, knew, or should have known, that this was not the case. The spring and early summer of 1911 had been among the hottest and driest on record. Small fires in the Porcupine Lake region and along the Minnesota border[4] had been burning months before Sutherland put pen to paper for his promotional articles in *The Globe*.

The government immediately went on the defensive when newspapers reported that there were supposed to have been a thousand men on the ground patrolling licenced lands, forest reserves, railway rights-of-way, and Crown territory. But there was little to defend when newspapers discovered that most of those jobs had not been filled because the pay was poor and seasonal and those that were had often been staffed by people with absolutely no training. An

LAND FOR SETTLERS

BY DONALD SUTHERLAND

DIRECTOR OF COLONIZATION.

THE people of old, or southern, Ontario have until recently regarded the northern part of the Province as somewhat similar to the highlands of central Ontario, valuable principally on account of its forests and the revenue to be derived therefrom. The prodigal manner in which vast areas were allowed to be devastated by the lumbermen and by fire, without any attempt at reproduction by conserving the small and growing timber, has left much of the rough land adjacent to the principal highway of commerce and Kenora. The Canadian Pacific Railway practically skirts the southern border of this vast territory for a distance of one thousand two hundred and eighty-five miles. Unfortunately the traveller over this road forms an entirely erroneous idea of the magnificent agricultural lands to be found a few miles to the north of the railway, and passes them by for the prairies of the west.

The building of the Temiskaming & Northern Ontario Railway from North Bay into this territory as far north as Cochrane, on the Grand Trunk Pacific, now under construction, has opened up and revealed a territory that for richness of soil is equal to any other part of Canada, and, as might be expected in a timbered country, the chance of crop failure through lack of sufficient rainfall is very small.

This vast area, estimated to contain from sixteen to twenty millions of acres of rich arable land, is situated in a most advantageous position. Its nearness to the seaboard, when compared with the western Provinces, is important. Cochrane, on the Grand Trunk Pacific, the terminus of the

5.3 Government of Ontario ad in *The Globe* and other newspapers in 1911.

editorial in the *Toronto Daily Star* complained that "the question of wages should not stand in the way in securing a competent fire-ranging staff." It lamented "the terrible loss of life" and demanded a "better system of fire protection in the North."[5]

Another editorial complained that there is a "growing conviction on the part of the public that fire-ranging is not efficient, that the laws are not enforced and the Government, while aware of the theoretical importance of conserving lives and timber in the northland, it is not prosecuting with sufficient energy the practical measures upon which conservation depends. The cry for help which comes from New Ontario today is a plea for future protection."[6]

Sutherland resigned a few weeks after the controversy died down, suffering no ill effect evidently from the publicity. He was elected to the House of Commons shortly after, assuming the seat for Oxford South until 1926, when he was defeated. He was appointed to the Senate in 1935 and held that job until his death in 1949.

The Matheson Fire of 1916

The Porcupine Fire should have been Canada's Big Burn, devastating enough to convince politicians to do whatever was necessary to make sure that it didn't happen again. But legislation and the strategies necessary to better manage fire were still not in place three years later when northern Ontario once again experienced a series of conflagrations.

Mercifully, no one was killed when the first of these tore through the town of Hearst, destroying as many as 500 homes and leaving almost everyone destitute. The town of Cochrane, however, wasn't as fortunate. Twelve people died when a matrix of fires tore through the area. The deadliest of these burned a hundred kilometres southeast in the region. Fire ripped through more than 200 million hectares of forest, levelling the villages of Matheson, Kelso, Val Gagne, Porquis Junction, Ranmore, and Iroquois Falls. More than 200 people died in the deadliest disaster since the Miramichi Fire of 1825.

It was big news, so big that even though the Allies had just made a tremendous assault on Germany's west front, the Matheson fire was the banner story in *The Globe* and other newspapers. There was really no other choice for editors when there were relief trains transporting dozens of doctors and nurses, food, clothing, and tents into a region where people reported seeing dead birds dropping from the sky. There were so many tragic tales to tell. Among the saddest was the death of ten children who died in their shack of a home after their mother went for help, not knowing that her husband was on the way. He died as well trying to get the children out of the burning building. Not far away, three men and five children were buried alive when the well where they took refuge caved in. Three women and another man who were not in the well were overwhelmed by smoke. Their bodies were found wrapped in once-wet blankets that had been baked by the heat.

This time, the government of Ontario could not ignore the fallout, although Howard Ferguson, the minister of Lands, Forests and Mines, tried to by wondering callously why many of the dead did not jump into a nearby creek to save themselves. He even went so far as to suggest that the amount of area burned was being exaggerated. Ferguson then redeemed himself by echoing what Gifford Pinchot and Henry Graves had been calling for south of the border. "The problem is to get together and concentrate your firefighting forces

5.4 With 223 dead, burying all the victims of the Great Fire of 1916 was a big job for the survivors. Photo from the book *Killer in the Bush: The Great Fires of Ontario* by Michael Barnes, 1987.

as soon as fire is discovered." Then he announced that Edmund Zavitz, the head of the forestry section, would lead a new branch of government to deal exclusively with forestry and fire prevention.

Schooled in forest management at Yale and the University Michigan, Zavitz was the first professional forester in Ontario and one of a new breed of forestry professionals across Canada. Earlier in his career, working for the Ontario government, he had seen how wildfire, agriculture, and clearcutting had turned much of the southern province into unproductive wasteland. He was determined to return trees to these barren lands, even going so far as offering saplings to every farmer. When that plan failed to be adopted by the government, he was put in charge of setting up and running a provincial forest station. It was the first of its kind in Canada.

Zavitz was well liked by his peers. "Fuzzy" was the nickname they gave him before his full head of hair thinned out, possibly from the stress that came with the job. His mandate to better prevent and suppress fire was supposed to have been made easier when the Forest Fires Prevention Act was passed the year after Matheson burned. The act put into place a permit system for burning in northern Ontario. A field fire organization was created and equipped with newly developed mechanized equipment. A network of fire detection towers, ranger stations, and equipment warehouses was built across forested areas of the province and linked by radio or telephone. More firefighting gangs were assigned to patrol the backroads in trucks and on hand cars along railway lines.

Yet, even with these new measures in place, the forests of Ontario were still not safe because bureaucrats in fire-prone areas were neither enforcing the rules nor hiring enough people. And many of those that were involved in fire detection had only canoes to get about in heavily forested landscapes, where paddling from one place to another was often impeded by strong winds, dangerous rapids, and boot-sucking, buggy portages. They might have smelled or seen smoke coming from a fire, but it would sometimes take a week to figure out where it was coming from.

5.5 Firefighters often used handcars in the early part of the nineteenth century. Bala, Ontario, 1916. Photo by John Boyd.

The situation was similar all across the country.

The war was partly to blame for the shortage of staff in Ontario and within the Dominion Forestry Branch, which was established in 1901. In 1916, thirty of the fifty-five rangers, clerks, engineers, and assistants who worked in the branch had gone to war. There was no plan to replace them until they returned. Even if substitutes had been found, noted Robert H. Campbell, the man who replaced Elihu Stewart as director in 1907, "the great extent of the country and the scarcity of population make the territory which each ranger patrols almost hopelessly large, while it is nearly always impossible to get sufficient help to cope with a fire of any size."[7]

That was an important point. Canada couldn't expect to be as good as the United States in managing its forests when it had more trees, only a tenth of the population, and an even smaller tax base from which to draw funding. There was more to the problem than that, however. Ernest Finlayson, district inspector of Forest Reserves in Alberta, spelled it out in a memorandum to Campbell. "Speaking generally with regard to the fire-ranging organization, I would point out that it will probably never be possible to develop a highly efficient fire-preventive organization under the present methods. Although the number of fires which are extinguished by the ranger staff at present provided justify, in potential dollars and cents, the expenditure made, it is certainly quite true that no highly specialized organization can be developed under the present system. The fundamental reason for this is that, when this branch has no administrative control over lands embraced within the fire-ranging districts except in so far as fuel protection is concerned, it is impracticable to consider the provision of adequate transportation facilities and other improvements, upon all of which any successful fire protective organization is absolutely dependent."[8]

In spite of this clear articulation of the complexity of the issue, Indigenous people continued to be seen as a big part of the problem. "The main sources of danger from fire in the north are travellers, including Indians," Campbell asserted, without providing proof. "Indians generally have the reputation of being careful with fire but the occurrence of many fires at a distance from the regular routes of travel and on trails followed only by the Indians would indicate that they have not yet learned to take the necessary precautions." In fact, Campbell was well aware that the railroad companies were the biggest problem. He had authored two reports, one in 1911 and another in 1916, that called attention to fires caused by railway companies.[9] He even predicted that the 123 fires that railway companies were responsible for in 1916 would be repeated in future years.[10]

Nor did Campbell offer solutions, at least not publicly. He left that up to Clifford Sifton and the Commission of Conservation. "There are still 4,087 miles of Dominion Government railways and 350 miles of provincial chartered railways in Alberta not subject to regulation and inspection by the Railway Commission, and the fire prevention service applicable to these two classes is not comparable to that applied under the Board of Railways Commissioners," Sifton wrote in the COC's annual report of 1917. "The result is that on the Government's own railway (The Intercolonial Colonial Railway of Canada), the fire protective service is the worst we have in Canada. Urgent necessity also exists for dealing with the northern railways in Alberta."[11]

Sifton was right. For such a small population, Canada was saturated with railroads. The maniacal drive to expand them into the hinterland continued without the checks and balances needed to prevent the big destructive wildfires that were routinely torching northern Ontario from extending their reach into the boreal forest. But that was pretty much the last Sifton had anything to say about what needed to be done. He resigned that year because of poor health, and possibly out of recognition that the COC's days were numbered.

Lac la Biche

In the summer of 1919, twenty-three Cree people were camped by the edge of a lake 300 kilometres east of Lac La Biche near the Saskatchewan border when a wall of flames came at them. Theresa Desjarlais was ten years old at the time. Years later, she recalled what happened when her father yelled "Fire," and pulled her out of the tent that afternoon.

> It was pitch dark but there was a yellow glow which seemed to reach to the sky ... balls of fire were falling all around us; the jack pine trees had become ignited and were like big torches. By the time we reached the lake the heat was terrific. Mother had managed to grab a blanket and a horsehide robe from the tent; these Father threw in the water and covered three of us with the blanket ... Firebrands were falling all around us. Some fell on our blanket and we had to keep it dipped in the water constantly to prevent it from burning over us ... We were all badly burned, especially my father. The horsehide which he had thrown over my mother and little sister had burned to a crisp on Mother's back.[12]

Theresa's father died the next day of injuries he sustained while keeping the protective blankets wet. The survivors spent two miserable nights without food or shelter. When help finally arrived, it took another five days of hard travel to get to safety. Thirteen other people died as well, including two Cree brothers who were on a hunting trip in another part of the province.

No one knows how those fires started. But it wasn't hard to guess. Just a year before, sparks from a train ignited wildfires in northeastern Minnesota, killing a thousand people and destroying the towns of Cloquet, Moose Lake, and several other communities.

The circumstances in Canada were eerily similar. Towards the end of 1918, the Canadian Northern Railway, a privately owned company that was built to compete with the CPR, had pushed through to Lloydminster, Battleford, and Prince Albert, Saskatchewan. A spur line from Prince Albert to Big River to the northeast served the sawmill industry as well as farmers, many of whom worked in lumber camps in the winter. Given what had previously happened in other parts of the country, there was no reason to believe that sparking trains, or possibly timbermen and farmers burning slash would not set the stage for the next big fire.

The winter of 1918–19 in Saskatchewan and Alberta was notable for the lack of snowfall. Spring brought little rain. The forest was tinder dry by the time the fire season arrived. Whether it was sparks from a locomotive, farmers burning their fields, or lumbermen torching slash piles, no one knows, because Dominion Forestry Branch rangers were nowhere to be found. As Finlayson noted, they

5.6 Toronto streetcars sent to Haileybury in 1922.

did not have clear authority on provincial land at the time. In addition, many of them were not on duty yet, after having just returned from fighting the war in Europe. What is known is that several clusters of fires ignited on both sides of the border between Alberta and Saskatchewan in May and June, surrounding towns like Prince Albert, Lac La Biche, Big River, and the First Nations reserve at Lac des Îles, east of Cold Lake where those Cree people were camped.

Had it not been for hundreds of lumbermen enlisted as firefighters on a moment's notice at Big Lake, death would have come to that region as well. Settlers in Lac La Biche would also have perished had not been for the foresight of town leaders who encouraged everyone to jump into the lake with soaking blankets on their heads to protect them from the heat and smoke.

There were so many fires burning in Canada that summer that the *The Globe*, the *Toronto Daily Star*, and other eastern papers barely noticed what was happening in Alberta and Saskatchewan. That was left to the *Edmonton Bulletin*, a local paper that first learned of the fires in east central part of the province when the settlers of Lac La Biche wandered into town with only the clothes on their back.

> The little village of Lac La Biche is today a mere smouldering mass of ruin and desolation, and its entire population is homeless and bereft of all personal effects, save scant articles of clothing which could be worn through the nerve-wracking struggle the people were forced to make to preserve their lives … The absence of a death toll in the catastrophe is due to the heroic measures taken by the citizens, who rushed into the waters of the lake and defied suffocating heat and smoke by means of wet blankets. Only such measures saved many of the women and little children, the intensity of the fire being shown by the burning of the very reeds along the shore and surface of the lake.[13]

All told, the fires in Saskatchewan and Alberta blackened more than 2.8 million hectares of forest, making it one of the largest complexes of fires in North American history. Robert Campbell was quick to blame settlers and their "unregulated use of fire" in clearing the land. The Dominion Forestry rangers could not act as efficiently as they might have, he contended, because the Prairie and Forest Act didn't give them the authority to do so.[14]

The economic damage, the loss of life, and the realization that not enough was being done about wildfire on the prairies spurred governments into action. British Columbia and the Dominion Forestry Branch began using planes to patrol the forest in the west the following year.[15]An agreement was put in place to give Dominion Forestry Rangers some authority on provincial lands. And when Prince Albert National Park was established a few years later to promote tourism (the fires having destroyed most of the merchantable timber), the Dominion Parks Service did what it had done pretty much everywhere else. It evicted Indigenous people and moved them to Montreal Lake.

Haileybury 1922

Notable among the provinces that did not take up the Royal Canadian Air Force's offer to patrol the forests for fire was Ontario. The government declined even after the government of Quebec and the St Maurice Fire Protection Association agreed to cover the cost of the pilots' pay and the maintenance of the planes. Edmund Zavitz, still the chief forester for the Ontario government, must have been mortified when it was left to him explain to newspaper reporters that it was too costly and impractical.

The progressive Ontario Farmers Union party, which was brought to power in 1919 with E.C. Drury as premier, may have been too distracted by other pressing matters to turn its attention to fire. As well as forestry and mining, it had inherited a complicated matrix of backroom dealings that had given certain companies access to Crown timber either at bargain prices or without having to make bids for additional lumber.[16]

In addition, some pulpwood exporters would pose as settlers to get access to timberlands the government offered to homesteaders for free, or for very little. Others staked claims to explore for minerals they knew were not there so that they could harvest and sell timber instead. That allowed these "timber pirates" to avoid paying Crown dues. Meanwhile, mining companies often burned bush to expose minerals in the ground – a practice that Zavitz tried unsuccessfully to stop.

Previous governments had not been above engaging in backroom deals such as that in which Colonel James Arthur Little, a pulp and paper owner, allegedly placed three $1,000 bills on the desk of George Howard Ferguson, the minister responsible for Lands, Forests and Mines from 1914 to 1919. Commissioners overseeing an inquiry into Ontario's forestry practices didn't believe Little when he denied the accusation and shrugged his shoulders when bank records that were presented to him as evidence. They concluded that "the companies have trespassed upon Government lands without shadow of right. Very large quantities of pulpwood were wrongfully obtained ... In our opinion all such proceedings should be taken by the Attorney-General for the punishment of those who have committed forgery and perjury, as the evidence transmitted here with may justify," the inquiry recommended.[17]

The recommendation fell on deaf ears. Little and his partner Walter H. Russell, a lawyer from Detroit, managed to escape justice. The bad publicity did not stop them. Little simply moved on, joining an American company to form Nipigon Fibre and Paper Mills Limited the next year.

The situation in northern Ontario was such a mess that the Hydro-Electric Power Commission of Ontario was forced to go to court to get the Nipigon Fibre and Paper Mills Limited to pay for more than $93,000 in electricity. The Spanish River Company went unpunished even when it was hundreds of thousands of dollars in arrears in fire protection dues. Witnesses testifying at that inquiry in April 1921 conceded that this system "had led to serious losses to the public Treasury."[18] The accounting was so bad that Albert Grigg, the former deputy minister of Lands, Forests and Mines, testified that the bookkeeping practices in the department he once ran would be "disgrace to a country store." Being a party that grew out of the farmers' movement, the Ontario Farmers Union evidently did not want to bring the hammer down on settlers who burned to clear the way for planting crops. Rather than ban them from burning, the Ontario government took out ads out in newspapers exhorting settlers as well as lumbermen, and railway companies campers and Indigenous people, to be careful about lighting fires.[19]

Whether it intended to be or not, the government of E.C. Drury was an absentee landlord that had lost control of land it was effectively renting out to forestry and mining industries. Many of those companies were still unwilling to do their part in dealing with increasingly frequent wildfires. Newspaper articles that reported on the fire season of 1921 gave evidence of this. One published in *The Globe* stands out from the rest because it was simply a list of fires that the Temiskaming & Northern Ontario Railway identified on one of its runs through the region.

Small fire at Mileage 66
Small fire at Bay Lake
Small fires at Upo Park, Thornlee, Earlton. McCool, Heaslip and Charlton
Small fires at Boston Creek, Rosgrove, Dene, Swastika, Kirkland Lake, and Sesekinka
Monteith – small fires all over country
Porquis Jet – Bad fire
Bad fires around d Porcupine Lake
Bad fire five miles upriver of Timmons
Cochrane, Several fires, mostly all under control and no danger at any point unless wind would rise
Several fires along the Canadian National Railways from Hearst to O'Brien. All beingwatched and under control
Connaught, Several fires in the country – not dangerous unless wind rises
Iroquois Falls, Fires on all sides; no immediate danger

Edmund Zavitz never went public with his concerns the way Pinchot would have, but his frustration was evident in the annual report he wrote for that year. "The outstanding feature of forest administration in the Province, as in all Eastern Canada, is an inability to control the losses from forest fires. The undertaking is so large and its bearing so important that the other phases of administrative work are comparatively minor matters."[20]

Clifton D. Howe, who replaced Bernhard Fernow as dean of Forestry at the University of Toronto in 1919, was quick to take this up as public issue. In an article he wrote for *The Globe* in the summer of 1922, he chided newspapers for "devoting so much ink to pulp and paper tariffs when they should be focusing on fire and the height of the country's trees, because the products of the trees of Canada contribute more wealth each year to the country than any other activity of nature or man, except the production of agricultural crops."[21]

Howe was an American familiar with Pinchot's and Graves's transformation of the US Forest Service into a truly national agency with a modicum of state support and co-operation. The spate of destructive fires in Canada horrified

5.7 The Great Fire of 1919 was a complex of many fires that burned homesteads, Indigenous hunting camps, timber berths, and the town of Lac La Biche in Alberta, pictured here.

him. He pointed out that just the year before, wildfire had destroyed 307,500 hectares of forest in Ontario, 251,000 hectares in Quebec, 40,000 hectares in the Maritime provinces, and more than 607,000 hectares in total. That, he added, was the official count, and not truly representative of what was going on in the so-called hinterland. The area burned in the past seventy-five years, he estimated, was twice the size of Ontario. "The wood using industries cannot continue to increase the wealth of the country by millions of dollars a year unless the occurrence of forest fires is reduced to a minimum," he warned.[22]

At some point, this laissez faire approach to forest management would lead to another disaster. It could have happened anywhere in 1922. British Columbia was experiencing one of its worst fire seasons in years. In August, there were more than 2,000 men fighting fires along Minnesota-Ontario border. At one point, the province of Quebec was reporting 60 to 100 fires that were out of control. More than 4,000 fires were recorded in Canada that year.[23]

Newspapers noticed.

"From Lakes to Ocean, Forests Are Aflame," screamed the front page of *The Globe*.[24] Newspaper editors marked the eleventh anniversary of the Porcupine Fire with front page reporting on that terrible disaster, reminding their readers that the town of South Porcupine burned down in eleven minutes. But it was the fires that ripped through the Haileybury region of northern Ontario in October of 1922 that caused the most grief. Forty-three people died, and more than 6,000 were left homeless. Eighteen townships suffered more than $8 million in damage.

That fire might well have been prevented. As the autumn approached, seasonal rangers asked permission to stay on the job so that they could monitor and restrict the amount of burning farmers did after their crops were harvested. Bureaucrats refused their request and told them to go home. Without this oversight, farmers took advantage of the situation. The small fires got out of control on 4 October when a light breeze morphed into hurricane-force winds.

The devastation that followed was just as horrific as that in 1911 and 1916. "There are screams, moans, cries, it is sad to see and hear!" wrote eyewitness Hilaire Damphousse, in an undated manuscript. "The Great Fire has left desolation everywhere. The morning after, eyes aching and red, the people look, desperate, the six inches of snow, darkened by the ashes, that fell overnight and that is covering the carcasses of the cattle and the remains of the houses. Desperate to the last point, no one can say a word without letting tears drop. It's back to square one."[25]

Some were fortunate. A man by the name of Eli Thrub rushed home from Toronto to see what had happened to his house and family in the town of Charlton, which had been levelled. He discovered that his was the only house left standing. His wife and children were inside, traumatized but unhurt. Fellow resident James Desire rushed to the hospital when he learned that it was burning. His wife was a patient. He carried her in his arms into a shallow portion of a nearby lake and waited there shivering until the danger had passed. Presumably, he and others got through the following nights with provisions that were sent in by the T. Eaton company.

It was easy to blame the Ontario government. Following the devastating fires that burned in Baudette–Rainy River in 1910, the Porcupine Lake region in 1911, Matheson in 1916, and Haileybury in 1922, Ontario premier E.C. Drury knew that he, as leader of the populist Union of Ontario Farmers party, had a public relations disaster on his hands when the *Toronto Daily Star* quoted one woman who suggested that there was no point criticizing the government for the early withdrawal of rangers. "They are not even here in the summer time," she quipped.[26]

Drury spent two days touring the region in person and reported: "Too much cannot be said in praise of the fortitude and helpfulness of the Northern people. Throughout the fire area, while the losses have been tremendous, there has been no despondency and I find everywhere a brave determination to rebuild on a better and surer basis."[27] This tone deafness presented itself once again when he suggested that money was tight and that victims should not expect too much from the government. Then Drury did what many Canadian politicians did when they dug themselves into a deep hole. He called for a royal commission to investigate the cause of the fire and to make recommendations on preventing future catastrophes.

E.P. Heaton, the fire marshal who headed the inquiry, did not offer much help when he tabled his report. Mindful perhaps of how difficult and important it was for settlers to clear the land, he did not recommend banning the practice of burning. Nor did he mention what the railroad companies might do to prevent fires. Instead, he suggested that townships hire fire marshals like him who would grant permits to light fires. He also recommended that strips of land be cleared around settlements and municipalities to "retard the progress of fire."

Drury's government adopted these recommendations. And bowing to public pressure, it purchased twenty surplus US Navy flying boats. This did not turn out as well as he had hoped. One of the flying boats could not get off the water after several attempts at takeoff in Port Arthur, with the media watching. Another crashed while trying to land on the Bay Street docks in Toronto. Drury's reputation never recovered. With the help of pulp and paper companies, George Howard Ferguson, the former minister responsible for Lands, Forests and Mines who allegedly accepted a bribe, led an all-out attack on Drury and his government, describing the royal commission that implicated him as "claptrap political conspiracy," Drury guilty of "political knavery," and his party as a bunch of "intellectual freaks who were projected into prominence by accident and who grew out of garbage."[28] The United Farmers fell from power in 1923. Drury went on to become sheriff and registrar of Simcoe County.

6

PAIUTE FORESTRY

In the half-century that followed the Big Burn, Henry Graves, William Greeley, Robert Stuart, Ferdinand Silcox, and their successors in the US Forest Service put into play several strategies to reduce fires. Graves believed that forests on critical watersheds should be owned by the public for their protective value. "Public forests serve, also, as centers of co-operation with private owners and as demonstration areas for the practice of forestry as well as furnishing their direct benefits in producing wood materials, as recreation grounds." There was no role for Indigenous people to play in these ecosystems. The way Graves saw it, preventing fire necessitated the exclusion of Indigenous people from federally run lands.

Removing Indigenous people from national parks and public forests was a strategy that was already in play by the time he came along. John Wesley Powell had championed the idea back in 1882 when he noted that, while an arid climate creates the conditions for fires, it was "Indians" (not white people) who were the chief source of ignition. For that reason, "The fire can then, be very greatly curtailed by the removal of Indians."[1] This was done in parks such as Grand Canyon, Yellowstone, and in Yosemite, where Indigenous people burned to clear underbrush, open pasture lands for deer, and to support the growth of woodland crops such as acorns from oak and sweet potato roots. The assault on tribal groups was by this time in full force, akin to the war on wolves and coyotes – animals which a report in the *New York Times* described as "thieves" and "good for nothing at all."

6.1 Blackfoot men burning grass in southern Alberta in 1911.

Most Americans were still smarting from Battle of Little Bighorn in which General George Armstrong made his last stand in June of 1876 and were apt to agree with this comparison. "An Indian is almost as bad as a coyote in the manner of diet," the *Times* opined with a tone of contempt and racism that was so common at the time. "He will eat anything that is eatable, and is exceedingly filthy in his habits and not particular as to the daily bill of fare.[2]

The legality of evicting Indigenous people from public lands in the United States was partially settled in 1896 when the Supreme Court (Ward vs Race Horse) ruled that the establishment of Yellowstone as a national park and the admission of Wyoming as a state was the legal foundation for preventing them from dwelling there. Their treaty right to hunt and dwell in the park was deemed to be temporary. (The case revolved around Race Horse, a member of the Bannock tribe that once inhabited the region year-round. He was jailed for hunting elk in the park.) The decision gave the government licence to ban other Indigenous people, such as the Blackfeet, from entering parks and public lands like Glacier National Park in Montana when it was established in 1916.

William Greeley, US Forest Service chief from 1920 to 1928, had no qualms about continuing this practice once he took over from Graves. Wildfire, he said, echoing his predecessors, was a "menace." Prescribed burning was "evil." He dismissed the strategy of light burning as "Paiute forestry,"[3] even though it was producing positive results in places like Georgia where it was referred to as "woods burning."

"We should no more permit an essentially destructive theory, like that of light burning, to nullify our efforts at real forest protection than we should permit the advertisement of sure cures for tuberculosis to do away with sanitary regulations of cities, the tuberculosis sanitaria, fresh air for patients, and other means employed by medical and hygienic science for combatting white plague."[4]

It wasn't just politics, business, and racism that was behind efforts to stop Indigenous burning. Greeley had the backing of prominent scientists such as Aldo Leopold. Forests couldn't be productive under light burning scenarios, according to Leopold, because they would kill most of the seedling trees necessary to replace the old stand. Leopold believed that light burning gradually reduced the vitality and productiveness of the forage and destroyed the humus in the soil necessary for rapid tree growth (that is, rapid lumber production, and germination of the seed). "Light-burning, by inflicting scars, abnormally increases the rots which destroy the lumber, and increases the resin which depreciates lumber grades and intensifies subsequent fires … in most cases at least, increases the destructive effects of wood-boring insects … In other words," he wrote in 1920, "you can maintain some forests under light-burning, but you cannot maintain them efficiently."[5]

The thinking was similar in Canada when the first national park was established at Banff in the Canadian Rockies in 1885.[6] George A. Stewart, the superintendent of what was initially known as "Rocky Mountains Park," had been a civil engineer and landscape architect until the time of his appointment, not a dedicated conservationist or forester like Leopold or Pinchot. He and officials with the Canadian Pacific Railroad were more interested in creating spas, luxury hotels, steam-powered yachts on the Bow River and Lake Minniwanka, and even a zoo for wealthy tourists riding in on trains, than on maintaining ecosystem integrity. To that end, Stewart and Howard Douglas, his successor, were dedicated to the idea of preventing fires that blackened the park's so-called "ornamental trees." He frowned on the idea of the Stoney Nakoda continuing to reside in the park. "It is of great importance that if possible the Indians should be excluded from the Park. Their destruction of the game and depredations among the ornamental trees make their too frequent visits to the Park a matter

6.2 Chief Sitting Eagle was one of many Nakoda Stoney people who were banned from hunting in Banff. The only time they could return was during Pow Wow and Banff days. Photo by Chris Lund, 1957.

of great concern," he wrote in his first annual report. "The sight of primitive Indians," he reasoned, "deters wealthy tourists from coming to the park," except, of course, when they were brought during "Banff Indian Days" with full traditional head dress, buckskin jackets, and moccasins, drumming and behaving in ways that were more quaint than threatening.[7] These patronizing events were inspired by "Indian Field Days "in Yosemite where Indigenous people were put on parade, and poorly rewarded for Best Indian Squaw, for Best Indian Warrior costumes, and for an Indian Baby Show. Impoverished as they were, they could not easily decline the offers.

Douglas had the firm support of sports hunters who loathed the idea of the Stoneys shooting game for food that they themselves would have otherwise hunted for trophies. "They have systematically slaughtered the game," Douglas wrote in 1903. "Let the line be drawn now, if we don't the game will be gone."[8] In 1902, game warden George "Kootenai" Brown complained that the Stoney Nakoda "kill more game in a week than all the sportsmen kill in year."

Once the process of dispatching the Stoneys began, Stewart embarked on a sprucing-up program, so to speak. In 1887, he ordered 20,000 trees from the American Northwest to address the "want of variety in our foliage," and the absence of young trees in some places.[9] While it is probable that the Stoneys were responsible for some of the burning in Banff, it's also clear that explorers and fur traders also accidentally set fires in the forest. One example occurred in 1880, when surveyor James Hector and his party failed to extinguish a campfire that got out of control and ended up burning "a large forest near Glacier Lake." But the most identifiable sources of fire were the railroad workers and the locomotives that took tourists to Banff. "It is a matter of regret that fires incidental to the railway construction have devastated much of the country in the vicinity of the railroad and have spoiled much of the wonderful beauty the environs of these mountains," noted government surveyor J.J. McArthur in 1886. Locomotives shot burning embers from their chimneys. Railroad workers often emptied ash pans filled with smouldering embers onto, or along the tracks. They piled up slash and ties and burned them, often in hot weather when it was easier to light.

Everyone knew that steam-powered locomotives were a problem in Banff, as they were across the continent. American historian Francis Parkman said as much in 1896 when he denounced railroads as "the chief instrument for spoiling the woods."[10] Not only did they set the stage for even more game to be lost to the "skin-hunters, dynamiters, and netters" who exploited them, they also brought fire.[11]

The Dominion Forestry Branch, which was heavily involved in the management of Banff before the Dominion Parks Service got its start in 1911, knew railway companies were a problem when officials there issued a report in 1910 noting that railways were "well up in the list of the causes of forest fires ... If they do not lead, they always follow close in the black array."[12] The author of another government report complained that the railroad companies had "no interest in or care for public property."[13]

And yet, the notion that Indigenous people were responsible for the biggest fires persisted in many circles. Some even believed that Indigenous people lit fires in order to drive white men off their land. It was easier, of course, to do something about "Indian burners," as they were called, than it was to strongarm powerful railroad companies like the CPR, which was given $25 million, ten million hectares of free land, and a twenty-year land tax deferral in 1880 to build a railroad across the country, including across land that legitimately belonged to Indigenous people.

The CPR and other railroad companies were often given a pass when it came to preventing fire. The burden of that was left to the Dominion forest rangers and national park guardians who patrolled the tracks in Banff with velocipedes. Velocipedes were crude horse-drawn carts that were not easy to manage in rough

6.3 Grande Cache Métis visiting Jasper sixty-two years after they were forced to leave the national park.

terrain. But the guardians often prevailed. In 1910, they put out fifteen train-spark fires along the 100-kilometre rail line that connected Banff to Canmore.[14]

When another Rocky Mountain national park was established at Jasper in 1911, several Métis families that had been well settled there since the early fur-trading days were asked to leave. All good people, by most accounts. But they were also wood and grass burners when it came to growing hay, oats, and other crops. For a new national park in which fire control and tourism would be a top priority, this would not do, unless of course it was the new Grand Trunk Railroad that passed through the park.

J.W. McLaggan, the chief forest ranger in Alberta, was sent in to appraise Métis land and negotiate a price for their eviction. They could go anywhere except a national park. One couple, Lewis and Suzanne Swift, a Métis woman, held out, allegedly keeping surveyors at bay with guns. They eventually sold to a Brit who had the ill-conceived idea of turning their place into a dude ranch.[15] But they ended staying when Swift successfully applied for homesteading rights and was made Jasper's first warden. One of the official roles he played was to seal and lock up the abandoned homes of his former neighbours.

There was no end to these evictions in Canada. W.N. Millar of the Dominion Forestry Branch would likely have evicted the Stoneys from the forest reserves he was responsible for in the southern Rockies had he had the power to do so. Indigenous and Métis people were not so lucky in Algonquin in 1893, when the Ontario government prohibited all hunting,[16] in the Parc des Laurentides when the Quebec government issued the same order, and in Stanley Park in Vancouver, where the Squamish village of Xwayxway had been for more than 3,000 years.[17] One notable exception was the Temagami region of Ontario, where the Teme-Augama Anishnabai were allowed to stay because the Department of Indian Affairs had recognized their entitlements. Federal officials even had a plan to have them do "the work of caring for and operating the territory," an activity that would be "profitable for them."[18]

But, as historian Bruce W. Hodgins and environmental lawyer Jamie Benidickson have so well documented, Ontario government officials refused, viewing the Teme-Augama Anishinaabe as squatters and troublemakers rather than rightful landowners. Assistant Commissioner of Crown Lands Aubrey White believed it might be "very dangerous and even disastrous" if this First Nation could manage the forest. His concern was that they might become "enemies and set fire to the forest." One Ontario official went so far as question their "right to fish and shoot in the forest." Another prohibited them from cutting wood for timber.[19]

These eviction policies extended to even the most remote regions of the country. Many Cree, Chipewyan, and Métis were all but evicted from Wood Buffalo when a national park was created there in 1922 to preserve the last remaining wild wood bison herds on the continent. Even though a treaty allowed native people to hunt and trap in the park, federal officials would only permit them to do so if they could prove that they were signatories to a treaty that gave them that right. Many did not have the education or resources to do so. Métis who had married a person other than those of Cree or Chipewyan heritage did not qualify, even if they themselves could provide proof of their heritage. Orders were given to ban forever any Indigenous person who was found guilty of lighting a fire or killing a bison. Some were fined, but it was doubtful they had the resources to pay anything close to the $200 penalty, as posters warned in English and Cree syllabics. The policy backfired however, because the suppression of fire resulted in the deterioration of bison habitat.[20] A maturing pine and spruce forest shades out the sun, making it difficult for grass, sedges, and other plants that bison consume to grow.

John Macoun, a botanist working for the Geological Survey of Canada between 1882 and 1912, was one of the few who doubted that Indigenous people were largely responsible for fires. He conducted the survey that helped establish the route of the Canadian Pacific Railway through the prairies. "A stump of a cigar dropped on the prairie is much more dangerous than an Indian fire," he said.[21] And yet, when it was a cigar or some other form of carelessness that started a fire,

Indigenous people under the age of sixty living within ten miles of a prairie fire, or fifteen miles of forest fire, were compelled, if ordered, to fight those fires.[22]

Another low point came in 1936, when members of the Keeseekoowenin Ojibway band were expelled from the newly created Riding Mountain National Park in Manitoba. As the men, women, and children left with their belongings for a reserve outside the park, they saw smoke rising from their houses and barns. Park wardens set the fires to make sure they would not come back.[23] It wasn't just about fire, as historian John Sandlos points out. The removal of the Keeseekoowenin Ojibway from a rich hunting and fishing ground was a means of bolstering the Canadian government's program of assimilating Indigenous people through immersion in the supposedly more "civilized" occupation of agriculture.[24]

7

FIRE SUPPRESSION

As Indigenous people were incrementally removed from national parks and prohibited from burning on public lands, authorities on both sides of the border focused on a more organized, institutionalized approach to suppressing fire. The Americans led the charge largely because they were better funded, more politically motivated, and about three decades ahead of Canadians in the government management of forests. And of course, far more people in the United States were living in areas where fire was a serious threat. Something had to be done.

Fire lookouts were built in those wilderness areas that offered a panoramic view. The first were nothing more than tents perched on a wooden floor. The cabins that followed were constructed with hand-hewn timber to keep costs down. They were not the safest places to be. Fred Schmelling, who was stationed in a small cab on top of a thirty-foot pole tower on Mount Hood, Oregon, in the 1920s, reported that the hut had been struck by lightning three times in one night. When asked what he did in response, Schmelling responded laconically.

> "The first time I got hit, I sat up. Then I just got to sleep when another bolt struck the house, so I sat up again. After a while, it got struck for the third time, and I could feel rain coming in, so I got up and saw that part of the roof had been torn off."
>
> "What did you do then Fred?"
>
> "I lay down again."[1]

7.1 Aerial photograph of the fire lookout in Estudillo Mountain, June 2013. Photo by Kari Greer.

Schmelling was lucky. More than a few lookouts burned down when lightning struck them. Several of the inhabitants lost their lives.

As time went by, the design of these lookouts became more elaborate, capable of taking a strike from lightning while offering the inhabitants more comfortable quarters. The first fire tower built in California by a railroad company in 1876 was aided by a "fire finder" invented in 1911 by forester William Bushnell Osborne Jr at his home workshop in his spare time. This was a rotating steel disc attached to sighting mechanisms to help observers pinpoint the exact geographic location of smoke.

On the ground, rangers were dispatched into the backcountry where they would live seasonally in tents or old hunting cabins, or new ones that were built especially for them. Fire roads and trails were carved out through the more remote forested areas so that they could ride their horses in and keep watch for fire and poachers. Some patrols were done by rail car.

7.2 "Railroad Speeder Patrol" equipped with firefighting tools by the US Forest Service, travelling along the An Joaquin and Eastern Railroad, California, in 1913. Photo by E.G. Dudley.

Many of the firefighting jobs in those days were seasonal and low paying. The daily rate in northern Ontario in 1910 was about $2.50. That didn't cover the cost of food or transportation.[2] Some of the seasonal firefighters in Canada tended to be "townies" – high school and university students from well-to-do families who had connections in government. "We hardly knew how to put a pot on a fire to boil, or how to take it off again without spilling or burning our fingers and blinding our eyes with smoke," recalled historian Arthur R.M. Lower, who spent four summers from 1910 to 1914 as a fire ranger in northern Ontario. "No one troubled to give us a word of instruction … Had the students received some instruction and training, they could have been much more efficient than they were. Thanks to a general sloppiness of administration, the Ontario government did not get much of a return for the money it used on the old-fire-ranging system."[3]

It was a free-for-all, in many ways. The only rules governing these camps were those forbidding the consumption of alcohol and talking during lunch in camps where newly arrived immigrants spoke several different languages other than English. Some rangers augmented their food supply by dynamiting rivers in order to cough up fish. Camp managers kept the men content by allowing convoys of canoes to come in with women like Boxcar Rosie and Moustached Mabel, whose standard fee was five dollars.

In western Canada, the Mounties had always been part of the equation, fighting fires when they could but mostly educating settlers about the perils of burning. With the help of the Hudson's Bay Company and the missionaries, they implored Indigenous people not to burn, often with a poster written in the syllabic alphabets of the Cree and Chipewyan.

In the US, rangers tended to be better equipped, thanks to Pinchot's systematic war on wildfire. Initially, they used "heliographs" to send messages by means of a mirror and the sun's rays. These were soon replaced by battery-operated, single-wire telephone systems (also known as tree phones). The phones were eventually replaced by two-way radios.

In extremely remote areas where fire lookouts and the use of horses were not practical, canoes, and occasionally large boats were used, especially in Canada where the Mounties were trained to deal with fires in remote areas.

The idea of using planes to detect fires was first proposed by the US Forest Service in 1909, just a few years after the Wright brothers famously flew six kilometres, four times, over a field outside of Kitty Hawk, North Carolina. But nothing was done to move the idea forward in the Service until 1915 when E.L. Scott of the Eldorado National Forest wondered aloud, and then suggested in a formal letter, whether aircraft or dirigibles might be used to fight forest fires. The letter made its way to the chief forester's office in Washington, DC. Officials there were surprised to discover that the Wisconsin Forestry Department had already embraced this thanks to race-car enthusiast Logan "Jack" Vilas, who refused pay when it was offered, asking instead for "many thanks." Vilas is

7.3 RCMP fire crew.

7.4 Mackenzie River fire patrol boat. Photo by Franklin Hugo Kitto.

credited for making what is believed to be the first forest fire patrol flight on 29 June 1915.

Canadians took note the following year when Minnesota's state foresters tried to get the US Navy to undertake patrols from its Duluth base. The navy wasn't interested, but Elwood Wilson, chief forester of the Laurentide Paper Mill of Grand-Mère, Quebec, was intrigued, believing that lookout towers and "slow-moving rangers" were not good enough to protect business interests. In 1917, the St Maurice Forest Protective Association, which he helped found, arranged for test flights to see whether the use of planes was feasible and effective in detecting fires. Then, in 1918, British Columbia contracted the Hoffar Motor Boat Company to build a flying boat. Initial test flights were promising. But when the engine cut out in mid-air before the plane crashed into a Vancouver house, the government reconsidered its options, did the math, and concluded that aerial detection was too expensive.

The use of planes to detect fires became more practicable once the First World War ended. Air force pilots did not have much to do and a large number of planes were sitting idle. Canada received a gift of approximately one hundred planes that the British government no longer needed. The Americans had a similar surplus. So, when the US secretary of agriculture wrote a letter to the secretary of war asking for an air patrol to help detect fires, the request was accepted.

7.5 Curtiss flying boat. Photo by William Arthur Steel.

In 1919, Wilson convinced the Canadian government to lend two of its Curtiss HS-2L flying boats to the company for fire patrol. Pilot Stuart Graham and his wife Marguerite (Madge), possibly the first female navigator in Canada if not the world, flew the first one, *La Vigilance*, 1,040 kilometres from Dartmouth with mechanic Bill Kahre to their new base at Grand-Mère-Lac-à-la-Tortue in Quebec. The ten-hour flight was the longest by a woman at the time, though newspapers only referred to her as a "wife." Madge was not only the navigator: according to her grandson Rob Corcoran she was the media spokesperson and the one who rigged a clothesline on board so that the crew could talk while flying.[4] Two weeks after the planes went into operation, Graham reported the first fire spotted from a plane in Canada. He and a colleague conducted a total of fifty-seven flights in 1919. Graham later joined the Royal Canadian Air Force as a test pilot. During his time there between 1926 and 1928, he invented an automatic view finder for aerial photography and a folding canoe for use with float planes.

Most pilots were not fond of flying these behemoths. The planes were not easy to handle, slow in gaining altitude, and tricky to land on water. They offered little margin for error in mountainous country where downdrafts were common. And they required a mechanic because something would inevitably go wrong. More often than not, the crew was forced to hike out of the bush.

7.6 Radios were used to send messages from fire towers to central dispatch.

Between 1924 and 1932, half of the twenty planes that the Ontario government bought crashed, burned, or were damaged. One pilot involved in some of the 1,325 flights conducted in 1924 lost his life.[5]

The Americans had better success – or so they claimed. Despite the slowness of these wood and canvas planes, the dangerous terrain, and the lack of airfields, 442 fires were detected in California during the 1919 wildfire season, which had been the most destructive ever in the state up until then.[6] The pilots flew an estimated 325,000 kilometres in 2,457 hours.[7] The head of the service was so impressed, he sent a letter to the Senate informing its members, with some degrees of exaggeration, that the "patrols have made the danger of forest fire holocaust almost disappear."[8]

Air to ground wireless radios were still in the developmental stage back then,[9] and those that were available were in short supply. So, once a fire was detected, the pilot would drop messages or markers to show ground firefighting crews where to go. These markers, in some cases, were nothing more than rolls

7.7 **Homing pigeons readied for patrol on a Canso aircraft in 1942.**

of toilet paper. In other cases, carrier pigeons were brought along and released in midair with messages identifying the location of fires.

The idea of using carrier pigeons might seem crazy. But during the First World War, both the US Army and the US Navy used pigeons to send messages in conditions that were too dangerous for more standard forms of communication. Dispatching the birds during wartime proved extremely successful. According to one account, "some of them returned full of shrapnel; others were forced to walk most of the way after being injured so seriously they were unable to fly, and it is on record that one pigeon flew more than 11,500 kilometres (from France to Indo-China) to get its message through. Incidentally, the first word of the

Dieppe Raid arrived in Britain in the tiny compartment attached to homing pigeon's leg."[10]

7.8 The Smokejumper program began in 1939 as an experiment in the Pacific Northwest region of the United States.

When the war ended, the army offered to sell the birds to the US Forest Service for fifty cents each. Forest Service officials were intrigued, according to historian Diane M. Smith. Following a successful experiment using pigeons on Oregon's Deschutes National Forest in 1919, District 6 Forester George Cecil, in Portland, Oregon, submitted a budget to the service's Washington Office for a national program of aerial patrols. His budget included funding for pilots, equipment, and other air-patrol-related expenses, and for constructing pigeon lofts, purchasing pigeon food, and paying pigeon tenders. "Cecil's request for approximately $60,000 to fund the year's aerial patrol and communications program does not appear to have been acted on," according to Smith. "However, experimental use of pigeons for fire communication continued in Idaho and Oregon."[11]

Pigeons were enlisted for the same purpose in Canada in 1920, when the Royal Canadian Air Force, in co-operation with the Air Board of Canada, first began using planes to detect fires. The RCAF, which loaned out its pilots, eventually settled on the Vickers Vedette and the de Havilland Moth. The float or "boat planes," as they were called back then, were much smaller, more versatile, and more reliable that the Curtiss HS-2L flying boat. Charles Roy Slemon, the Canadian pilot who would eventually become the first deputy commander in chief of NORAD (the North American Air Defense Command), began his career in the RCAF as a fire detection pilot in 1924.

In 1925, John Weeks, the US Secretary of War, ended the program that allowed air force pilots to be involved in fire detection. The patrols were subsequently contracted out to the private sector. Interest in aviation did not wane in the agency, however. In the hopes of transporting men to a fire more expeditiously, the Forest Service considered using blimps and autogyros – a type of rotorcraft that uses an unpowered rotor in free autorotation to develop lift – before settling on parachuting smokejumpers into a fire.

The Soviet Union had come up with the idea of parachuting into fires in the early 1930s. Avialesookhrana (avian forestry), the Russian aerial firefighting agency (which now has 4,000 smokejumpers), originally trained men to crawl onto the wing of a plane, jump out over a village, and rally men to fight nearby fires. Legend has it that the first flight turned out to be an embarrassment when the pilot flew straight to Estonia in order to defect.

The US "smokejumper" was the brainchild of T.V. Pearson, the Forest Service intermountain regional forester. Officials initially deemed it to be too dangerous to drop men from the skies to slow or stop a raging inferno. For a time, they decided instead on parachuting cargo and equipment to help and feed firefighters on the ground.

Pearson, however, was persistent, shrugging at the amount of paperwork that was needed to get the experiments going, while dutifully putting pen to paper to answer questions from the airways inspector in the Department of Commerce who couldn't quite understand what Pearson was planning. He did his best to make it simple in a letter dated 9 April 1935:

> We desire to deposit men from an airplane to the ground by the method described, using the old exhibition of parachute. We desire to deposit men from an airplane to the ground by the method described, using the modern type of parachute. We desire to deposit men from an airplane to the ground, as stated in paragraphs one and two, using in each case only the one parachute loaded in the mechanism described, and without the man involved wearing a backpack chute as an extra-safety measure. In our experiments, the wearing of extra chutes constitutes only a handicap.
>
> Yours very truly
> T.V Pearson (acting) regional forester[12]

7.9 Ferdinand Silcox, fifth chief of the US Forest Service (1882–1939).

The plan to use smokejumpers prevailed. In 1935, thirty successful parachute drops were made in Colorado, twenty-eight using 155-pound iron weights and two involving humans.[13] In 1939, nine parachuters successfully completed sixty jumps. Their actions set the stage for the Parachute Project to go into full swing in 1940, three years after Ferdinand Silcox, by then the US Forest Service chief, came up with the policy which specified that all fires were to be brought under control by 10 a.m. the morning after their discovery.

Twenty years after the Big Burn, the United States finally had the tools it needed to supress fire. What it also had was more firefighters to use them than it had ever had before.

8

THE CIVILIAN CONSERVATION CORPS

In 1934, a wildfire burned through 73,000 hectares of historically significant land in the Lochsa and Selway river drainages of north central Idaho. This was the wild mountainous terrain that the Lewis and Clark expedition had travelled through more than a century earlier. It was the year that Ferdinand Silcox issued the directive requiring the US Forest Service to put out a fire by 10 a.m. the day following discovery. It was also the first of many years in which the US Forest Service had more men fighting fires on the ground than there had ever been before or would ever be again.

To provide jobs for the millions of unemployed during the Great Depression, President Franklin Delano Roosevelt signed an executive order that created the Civilian Conservation Corps (ccc), a New Deal welfare program which put three million men, not women, to work on various conservation projects. "She-She-She" Camps for unemployed women were set up as a counterpart to the ccc in 1933. Some called the ccc "Roosevelt's Tree Army." By 1 July 1933, 1,433 work camps had been established across the United State to plant trees, clear access roads, and fight fires.

The men came from all walks of life. Actors Raymond Burr and Walter Matthau worked for the ccc in Montana and California long before they became famous. Test pilot Chuck Yeager was involved, as was baseball great Stan Musial. The naturalist Aldo Leopold oversaw some of the restoration work done out of ccc camps in California. All told, ccc efforts represented 6.5 million days of fighting fires.

8.1 Civilian Conservation Corps fire crew.

James J. McEntee, the head of the CCC, was passionate about the program, seeing it not just as a form of welfare but as work that would promote tolerance. "Living together in barracks with other young men has a profound effect in teaching enrollees to respect the rights of others and to be tolerant of the ideas and beliefs of others. It teaches them to be good sports and to take minor defeats without flinching. These are traits of character which they will find exceedingly valuable in later life."[1] When he signed the executive order, Roosevelt made it clear that race and colour should not be barriers to admission in the program. Indigenous people and people of colour, however had no place in this cadre of brotherly love. Most of the 88,000 Indigenous men who worked for the Corps did so on their own reserves. African Americans were vastly underrepresented because the CCC didn't think it "practical" to "put colored enrollees into an area where there is no colored community."[2]

Fighting fires was undertaken in the traditional manner, with members patrolling the forests by truck, horse, on foot, and in canoes. When a big fire put people and infrastructure in harm's way, as it did near Los Angeles in 1934, 1,400 men were quickly dispatched with hoes, axes, pumps, and a bulldozer.[3]

The Corps worked to manage fire in other ways as well. In 1936 alone, CCC enrolees installed 72,000 kilometres of telephone lines, built 611 lookout towers, and cleared 18,300 kilometres of truck trails through national parks and forest. They maintained 112,500 kilometres of trails. Even a board game was created in association with the CCC to raise public awareness about the dangers of forest fires.

The CCC had its share of setbacks as well. In August 1937, fifteen firefighters, many of them from the CCC, were killed and thirty-eight more were injured when the Blackwater wildfire in the Shoshoni National Forest in Wyoming overwhelmed and trapped them on a rock ledge. Earl Davis, one of the survivors, described how "spot fires" suddenly popped up, penning them in. "For a time, we just sat there, hopeful the wind would take the searing flames away from us. As we watched the round ball of fire advance we thought we were gone. We lay down among the rocks with our faces to the ground. We couldn't get out. We could hear the fire approach – its roar was awful – and its flames licked out and scorched our backs like a hot iron. I know we felt that we were all washed up. Then the flames roared over us. We had to wait three hours for the ground to cool off so that we could walk to safety and help."[4] Davis speculated that it was panic that drove some of the men to run into the fire as it advanced.

An estimated 1,500 fires were burning in the West that month, prompting the *New York Times* to run the headline: "The Red Demon That Eats Forest."

8.2 Game board created in association with the Civilian Conservation Corps.

Add in the news story of an orphan dog named "Miss Hulda," a natural-born smoke chaser who went missing in southwest Washington around the same time, Congress could not look away and reluctantly agreed to continue funding the ccc, so long as the enrolment did not exceed 800,000. It had already rejected Roosevelt's proposal to make the program permanent. When ccc director Robert Fechner died in 1939, the days of the program were numbered. Its legacy, however, was a lasting one. Over a six-year period it had employed 2,600,000 men who planted almost two billion trees, constructed 172,000 kilometres of truck trails, erected 118,000 kilometres of telephone lines, and reduced the fire hazard to 809 million hectares of forest. More than nine billion person days were spent fighting fires.[5] The inherent racism which operated within its ranks, however, tarnished its accomplishments.

When the Second World War finally put an end to the ccc, a certain degree of panic set in because, in many places, the Corps had been first to respond to a fire. Claude R. Wickard, the US secretary of agriculture, testified before Congress in 1943 that the ccc had reduced the area burned by fire in the last eight years by about 20 per cent, or 3.6 million hectares.[6] At that same session, Congressman Malcolm Tarver of Tennessee wondered whether an enormous fire like the one in the Chattahoochee National Forest in 1943 was typical of what might happen in the rest of the country without the ccc. McEntee advised that it was, and that things would likely get worse, especially in the west. Earl Loveridge, assistant chief of the Forest Service and the mastermind behind the 10 a.m. directive, concurred, and warned Congress that without adequate forest fire capabilities, lands surrounding strategic facilities critical to the war effort would be in peril.[7] Defence officials such as Lt General John L. DeWitt were also concerned that smoke from wildfires had already reduced visibility for enemy aircraft warning stations, airplane training hours, and anti-aircraft training.[8] Loveridge was not alone in declaring that "there is no doubt that our enemies will burn our forests if they can. The fact that our forests are exceedingly more flammable adds to the tremendous risk involved."[9] DeWitt was of the same mind. After the attack on Pearl Harbor in 1941, he used his power to intern 110,000 Japanese women, men, and children, most of whom were American citizens, believing they would sabotage the war effort.

Following the bombing of Pearl Harbor, there was also genuine fear that the Japanese military might use their naval forces and planes to drop firebombs into coastal forests as a means of diverting resources away from the war effort. Without the ccc to suppress them, there was a real possibility that wildfires, whether they were started by the Japanese, Americans on the ground, or by lightning, would get out of control. To encourage American citizens to be vigilante, the War Advertising Council was created three weeks after Pearl Harbor by a group of advertising executives to bring members of their industry

8.3 US Forest Service wartime fire prevention advertisement.

together in the service of social good. These were the men behind the "Buy War Bonds" campaign, the "Women in War Jobs" push, and the forest fire prevention initiative that was managed in partnership with the US Forest Service.

The results were clever, artistic works of menacing genius. One of the US government posters they came up with had an image of Hitler and a snarling Japanese prime minister Hideki Tōjō together in the foreground of a forest fire. "OUR CARELESSNESS, Their Secret Weapon," the poster warned. Another was overtly racist in depicting a highly caricatured Japanese soldier grinning sinisterly behind a lighted match.

The Japanese were smarter than the Americans had assumed. A group of their military scientists came up with the idea of floating fire balloons on the jet stream to be transported 8,000 kilometres across the Pacific Ocean. The balloons had triggering devices that would drop firebombs over the forests of western Canada and the United States. More than 9,000 paper balloon bombs were launched in the hopes they would ignite so many fires that the US government

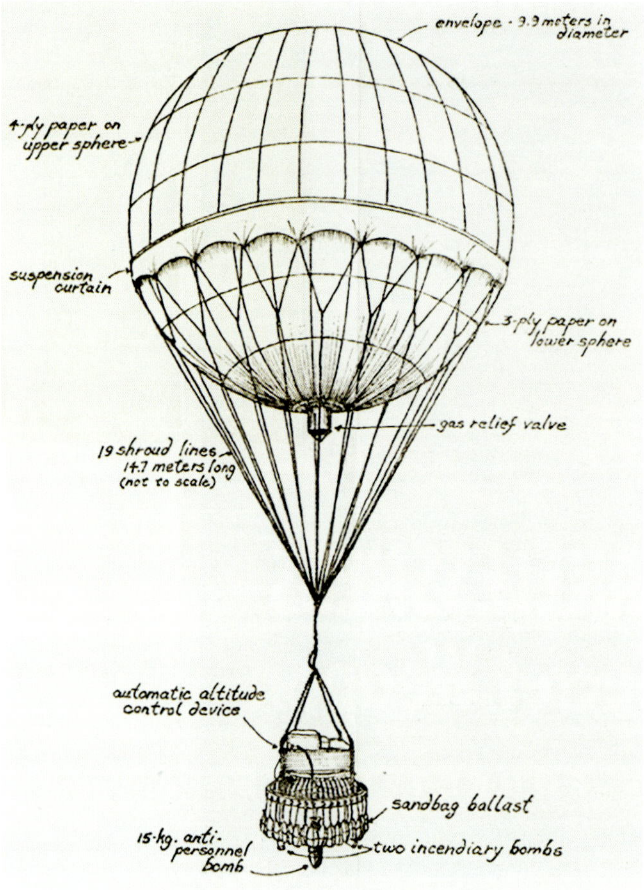

8.4 Sightings of Japanese airborne bombs began cropping up throughout the western US in late 1944.

would be compelled to withdraw men, planes, and boats from the war effort to extinguish them. It didn't work. Only 1,000 of the balloons made it to North America. One got as far as Grand Rapids, Michigan. Another got within ten miles of Detroit. Most landed in the ocean.

They got close, though. One airborne bomb damaged a generator at the Hanford Engineer Works reactor in Washington State, where plutonium was being processed for the first atomic bombs. An antipersonnel fragmentation bomb in southern Oregon killed a pregnant woman and five children.

Operation Fu-Go (an acronym which means "a wind ship weapon")[10] was seriously flawed in that the Japanese scientists failed to appreciate that the forests of the Pacific Northwest are typically wet in the fall months, when the balloons were dispatched. In some places, there was already snow on the ground. Nevertheless, sightings of the balloons in the air and unexploded devices on the ground created panic. When news of the Japanese balloons came to the fore, the military responded by sending out a group of African American paratroopers

8.5 The US government placed ads in newspapers and magazines in which wildfire was equated with enemy forces. Painting by Willy Pogany.

to the northwest coast to learn how to defuse the bombs and to fight any fires that might be ignited. The paratroopers were nicknamed the "Triple Nickle" in recognition of their being part of the 555th Parachute Infantry Battalion. Trained by the Forest Service, the plan was to drop the men from C-47 transport planes.

Officials from the FBI, the Army, and the Navy were also sent to figure out what had happened. American news organizations were urged to keep the operation secret. Most news organizations were more than willing to go along with this request. They had already been enlisted in the war effort and were happy to be paid to run full-colour advertisements demonizing fire and claiming that the conflagrations "delayed victory." It wasn't just the Japanese they were targeting; it was campers, smokers, industry, and anyone who was carelessness enough to start a fire.

For their Forest Fire Prevention campaign, the Forest Service and the War Advertising Council relied on a pool of famous or soon to be famous artists such as Norman Rockwell, Walt Disney, and James Montgomery Flagg (who

8.6 "Smokey Bear" was born on 9 August 1944 as the symbol for the US Forest Service and the Ad Council's joint effort to promote forest fire prevention. Painting by Albert Staehle.

was responsible for the Uncle Sam "I Want You for U.S. Army" poster). Under contract, the artists were paid $300 for each design. Like the ads with Hitler and the Japanese prime minister, there was nothing subtle about the message aimed at the American people. In *The American Weekly*, a magazine that claimed to have the greatest circulation in the world, one poster depicted a man with an axe about to strike down a wolf that has burst into flames in the middle of a forest. The ad ran with a message from Claude R. Wickard, the US secretary of agriculture, asking readers to be careful. "Our forests are vital to victory."

Another article in *Victory*, the official weekly bulletin of the Office of War Information, described man-made fires as "one of the Nation's most insidious enemies," and a "perpetual foe of conservation that is now sabotaging an important part of our war program." "Until we smash the Axis," the article quoted Wickard as saying, "forest fires are enemy fires."[11]

The most famous of these posters was the one of Smokey Bear. The idea was conceived by the Forest Service and the Ad Council in the dying days of the war. They initially considered using an owl, squirrel, or chipmunk before settling on a

8.7 Walt Disney's Bambi was also used in forest fire prevention posters.

bear for the fire prevention campaign. Artist Albert Staehle of *Saturday Evening Post* fame wasn't given a lot of leeway. The committee called for a pine forest setting, and a panda-type bear that was either black or brown. It had to have a face that was expressive, appealing, knowledgeable, and quizzical. The bear was to wear a hat like that worn by Boy Scouts.

The success of the poster exceeded even the most optimistic expectations. Smokey Bear was second in popularity only to Bambi, who was the first choice for the fire prevention campaign. The motto that was ultimately decided on – "Only You Can Prevent Forest Fires" – resonated in such a way that the US Postal service issued Smokey his own postal code to accommodate the thousands of letters that were sent by children each day. Smokey appeared on lunch boxes, toys, and pajamas. The licensing revenues went to the Forest Service.

The firefighting capacity of the Forest Service was, by this time, essentially too big to fail or to be subjected to the theory that fighting small fires made conditions worse because it left more fuel on the forest floor. As Stephen J. Pyne noted, these battles that started with Pinchot only set the stage for bigger defeats.

9

CANADA'S CONSERVATION CORPS

The irony in Congress passing the act that provided funding for Roosevelt's Tree Army in 1933 is that the Civilian Conservation Corps (CCC) was inspired in part by the Depression-era work camps that the Canadian government set up in 1932. Andrew McNaughton, the chief of the general staff at the time, was the mastermind behind these camps. Touring the country's military bases that summer, McNaughton was horrified by the number of people he saw sleeping in the streets and bunched up in open railcars looking for a job in the next town. What concerned him most was the possibility that widespread unemployment could result in a revolution.

Conservative Prime Minister R.B. Bennett was uneasy but ultimately receptive when McNaughton came to him with a plan that would get an estimated 70,000 men off the streets and railcars into relief camps. By the time the CCC was established south of the border, there were so many unemployed single men working in national parks such as Riding Mountain in Manitoba, Price Albert in Saskatchewan, Elk Island in Alberta, and in Waterton along the Alberta-Montana border that the government had trouble providing them with food and shelter. Minister of the Interior Thomas G. Murphy promised that many more would be employed in clearing brush and building roads like the one that eventually connected Jasper and Banff. Provinces such as Ontario bought in, seeing opportunity to expand the Trans-Canada Highway system in the north. By the time the program shut down, more than 170,000 men had stayed in 200 camps in every region of the country.

9.1 Strikers from unemployment relief camps en route to Eastern Canada during the "March on Ottawa," 3 June 1935.

It was supposed to be a civilian operation, but a military mindset prevailed. Canadian expeditionary sergeants in the Department of National Defence ran the show. Union members were banned from enrolling. Those that were accepted worked for twenty cents a day, a fifth of what the CCC program and She-She-She paid in the US and less than half of what American prisoners were paid when they were employed to fight fires. Instead of planting trees, fighting fires, clearing trails, and building fire lookouts, the relief workers did little of public value. Living conditions only got worse over time, as J.S. Woodsworth pointed out when he read portions of a letter smuggled out of a camp to fellow members of the House of Commons:

> Picture to yourself a tarpaper shack 79 feet × 24 with no windows, along each side there is a row of double decker bunks, these are spaced off with 8 × 1 board so that there is room for two men in each bunk. The bunks are filled with straw and you crawl into them from the foot end. Along the front of the lower bunk a narrow board is placed upon which the men may sit. The place is very meagrely lighted and ventilation by three skylights ... So narrow is the passageway between the bunks that when the men are sitting on the bench there is scarcely room to pass between them. This shack houses 88 men ... At times the place reeks of the foul smell and at night the air is simply fetid. The floor is dirty and the end of the shack where the men wash ... is caked with black mud. The toilet is thoroughly filthy, unsanitary, and far too small.

Rather than being immunized from socialist propaganda, as the government had hoped by locating the camps in the wilderness, thousands of relief camp workers ended up as members of the Communist party. In 1935, many of them joined others in riding the rails east to Ottawa to protest low wages and poor living conditions. Oblivious to the fact that the much of the public was rooting for them, police in Regina moved in as the relief camp workers were staging a peaceful protest in the town's square. Two people died, hundreds were injured, and untold damage was done to the city. Canadians, as historian Desmond Morton observed many years later, "chose (Mackenzie) King, not chaos in the election that fall."[1]

It was not only a terrible misreading of public opinion; it was a lost opportunity for better forest management. Support for a program akin to the CCC came from many quarters. *The Globe* pointed out in 1933 that Canada would do well if it created something similar to hire 817,000 unemployed people to better manage the forests.[2]

A number of people in the forestry industry felt the same way. Notable among them was Frank Barnjum, a Montreal-born businessman who was heavily involved in the trade of pulpwood between Canada and the United States from 1915 until his death in 1933. The son of British immigrants, he worked his way up in life, first as an office boy in a Montreal stock brokerage, then as tanning bark purchaser when he moved to Boston. He went on to be a horse-breeder, real estate developer, and land speculator, purchasing large areas of forest in Maine, which he exploited to produce pulpwood. By 1915, he was the owner of one of the largest pulpwood holdings in North America.

Barnjum wasn't done adding to that wealth just then. When he realized that some of the Cape Breton forests he had leased were too expensive to cut, he sold his interest to the Oxford Paper Company of Maine, owned by Hugh Chisholm, whose Canadian father had also moved to Maine to make his fortune

9.2 Frank Barnjum (1858–1933) is pictured here on Vancouver Island, British Columbia, two years before he died.

in the industry before he died in 1912. It was a sweet deal for Barnjum and an unsavoury one for Chisholm. When the company hired Bernhard Fernow, the dean of forestry at the University of Toronto, to survey the property, Fernow found that Barnjum had grossly exaggerated the amount of wood there was to harvest.

Barnjum's views on conservation unfolded in the same way they did with Gifford Pinchot. It wasn't owls and clearcuts he was concerned about. Like Pinchot, who came from a wealthy family that had laid waste to the forests they owned or leased, Barnjum realized that the resource was finite and that someday there would be no trees to cut down for commercial purposes because whatever was left would be claimed by fire. The embargo he proposed on the export of pulpwood to the US would ensure that his companies had enough trees to expand. Some critics went so far as to call him a "snake oil salesman" and a medicine man. A more generous assessment referred to him as an "opportunist par excellence."[3]

9.3 Cartoon mocking Frank Barnjum, 1925.

But the newspaper-reading public knew nothing about his profit motives and Barnjum was a bulldog when it came to promoting forest conservation. He donated tens of thousands of dollars to the Canadian Forestry Association, regularly took ads out in newspapers, sent hundreds of letters to newspaper editors, and made sure that every member of Parliament, as well as several provincial legislative assemblies, received a copy of one of the pamphlets he published between 1920 and 1923. The message was always the same. Too many trees were being cut down and too many forest fires were crippling the forest industry. "The apathy of our governments in connection with the serious forest situation is nothing short of criminal," he wrote in a typical letter to *The Globe*. "They tell their officials not to alarm the people with regard to forest fires that are eating up our wooded areas week in and week out; in other words, if a man's house is on fire, do not wake him up, let him burn."[4]

Barnjum's reach was national as much as it was provincial. He sponsored two essay contests on how best to protect the country's forests from fire and insects. The winner of the first essay contest was Otto Schierbeck, a Danish forester who worked for Price Bros in Quebec City, and who saw fire as the biggest threat to the nation's forests. Schierbeck went on to become the chief forester for the province of Nova Scotia, a position which Barnjum had lobbied for both before and while he was a member of the legislature. Schierbeck immediately went to work to promote better forest practices and a professional forest ranger service whose most important job was to prevent and supress fires.[5]

As British Columbia submitted to the clearcutting of vast stands of old growth forest, Barnjum stepped in and purchased 5,179 square kilometres before it was, as he said, "too late" to save it. Critics suggests that the trust he established to save the old growth forest was actually a limited company, "cannily organized" so that all its shares were held by his son, daughters, his estranged wife, and other relatives.[6] Still, Barnjum was too loud, too powerful, and too savvy to be ignored or successfully scrutinized, especially when it came to warning Canadians not to make the same mistakes American made across the border. "I have seen what happened to the forest in the United States, and the forest history is repeating itself in Canada only at a much more rapid rate in 1930."[7]

It all sounded so credible coming from a rich successful Canadian who had lived a good part of his life in and outside of Boston. Canadians loved fellow countrymen who made it big in the US. His complaints led to a royal commission on the pulp and paper industry, the reforms that Schierbeck instituted in Nova Scotia, and a fractious debate about how forests were being laid waste in British Columbia

Ernest Finlayson, acting director of the Dominion Forest Service in 1922, was torn by the much-needed publicity that Barnjum brought to his campaign for forest management reform and the hype that threatened to discredit it. "The propaganda carried on by Mr. Frank Barnjum is the most extreme type," he stated in a letter to Deputy Minister of the Interior W.W. Cory. "While this is it pleasing indeed to Canadian foresters to find a man of Mr. Barnjum's standing to undertake and finance conservation propaganda, it is felt that some of the data presented by him can hardly be justified." Then he added a note of caution: "Mr. Barnjum is a very wise advertiser, and at any time liable to make free use of communications he may receive, the full political and strategical significance of any letter sent him, therefore, might well be subjected to careful analysis."[8]

Politicians knew that it was better to back Barnjum than to attack him, as the pulp and paper industry tried to do during and after hearings of the royal commission on the pulpwood industry in 1923. The Canadian Pulpwood Association, in particular, went after him with every resource they had, taking out ads, distributing pamphlets, and publishing broadsheets that suggested, of all things, that he was the "dupe" "of pulp and paper companies out to increase profits at settlers expense."[9]

Barnjum's public image was somewhat tarnished by these assaults, as well as his refusal to testify before the commission. Finlayson made note of that in the commission's report of 1924. But Barnjum was always one step ahead of his enemies, crossing the border at one point to avoid being subpoenaed by the commission, and making his case directly to the newspapers. Most magazine and newspapers editors had his back. *Saturday Night* described the commission as a "farce" that was "dominated by American interests and operating solely to establish their contentions as to how Canada's pulpwood resources shall be exploited."[10] The *Globe*, which had raised concerns about fire and forest management for decades, barely noticed when the 292-page report was tabled, devoting only five paragraphs to it on page 3.[11]

It's no surprise, then, that the government that despised Barnjum appointed him to help form the Canadian delegation to the forestry committee at the Imperial Conference of 1930. Better a friend than enemy, the logic must have been. This gave Barnjum an opening to travel to China, Japan, and South America to gather information on forest conservation. One of his last trips was to Rome to interview Mussolini, whose government was in the process of draining the Pontine Marshes. Barnjum was such a high-profile figure by then that in January 1933, *The Globe* published his six-point plan to save the country's forest. Full stumpage rates, he wrote, should be charged on all fire-killed wood. Slash disposal regulations should be instituted to prevent future fires. More money was needed for fire prevention. The provinces and federal governments should buy up more forests to stabilize climate and protect watersheds. More money was needed for reforestation. Reductions were needed on the amount of wood cut on Crown land.[12]

When Barnjum died in Paris a few weeks later, *The Globe* called him a "doughty champion of this country's dwindling timber wealth."[13] The *New York Times* praised him for opposing the "spending of large amounts for forest fighting without equal sums for forest fire prevention."[14] The *Wall Street Journal* said he was best known in Canada as a "tree crusader."[15] More than few people suggested naming a national or municipal park in his honour. His legacy lived on for a time.

Conflagrations of the late 1920s and 1930s are rarely included in the lists of big fires in Canada. It is difficult to know if things were as bad then as Barnjum claimed. A national organization keeping track of what was happening across the country did not come into effect for some time because of Prime Minster Mackenzie King's decision to turn control of the western forests over to the provinces in 1930. Alberta was the first to begin recording in 1931. Other provinces took longer to do so.

Thanks to newspaper coverage and local history books, we know that fires burned so often in the 1920s and 1930s that, at one point, *The Globe* wondered whether it was best to simply permit fires to burn themselves out and fight only those that put people and infrastructure in harm's way.[16] There were many reasons why fires burned so big and so often. In 1933, the Pacific Science Congress reported that in the manufacture of twelve billion feet, board measure

9.4 Residents of Cranberry Portage in Manitoba fleeing a wildfire that destroyed their town. Photo by Williams.

of lumber (a "board foot" is a piece of lumber twelve inches wide by one inch long by one inch thick), ten million cords of white wood (a "cord" is 128 cubic feet of wood) that could be turned to pulp were left to rot or burn. And where British Columbia had logged 51,000 hectares annually from 1922 to 1932, an additional 233,000 hectares burned, mainly because of the careless practices of the lumber industry that allowed slash piles to ignite.[17]

It wasn't just a BC problem. Sixty percent of the logged forests across the country ended up as slash. In Ontario, where waste and carelessness were just as bad or worse, the government was spending $1,200 or more than it had allocated in its budget fighting fires throughout the 365 days of 1933 – an inexcusable waste, according to *The Globe*.[18] The Alberta Forest Service was so overwhelmed fighting fires around the Sibbald fen in the Kananaskis area of southwestern Alberta in 1936 that officials appealed to the Dominion Forest Service for help.[19] None came.

Most of the work was left to local farmers and ranchers who used ploughs and tractors to dig trenches in the path of the fire. But these fire guards were no match for ninety-mile-per-hour winds that gusted with such velocity they lifted the front end of a newly built barn. "The wind was so strong, men could only walk by hanging onto fences and little children were blown around like piece of firewood," recalled Percy Copithorne, who was there to witness it.[20] Smoke was so thick that chickens were going to roost in the afternoon and children were "accepting the unnatural blackness as an early curfew."[21] Burning fences posts were the only guides for people who tried to drive to safety along road allowances that were shrouded in thick smoke. Fourteen children and a teacher would have perished had it not been for two brothers who rushed to the school through the blinding smoke and dust to pile them into their cars. One man who suffered from pleurisy died after he was taken to hospital in a state of shock. At least twenty families lost their homes, barns, granaries, garages, and livestock. The carcasses of pigs and cattle were strewn about everywhere.

News of the fire attracted hundreds of sightseers who drove in from Calgary and other towns. The *Calgary Herald* launched a fundraising drive to help the destitute. Even when the rains finally came, the peat in Sibbald fen smouldered all winter long, lighting up at times as if it was being illuminated by hundreds of little candles.

Notable among the other conflagrations in the thirties was the MacBeth Fire near Sudbury in May 1932. It resulted in a 1935 Supreme Court decision favouring the insurance industry over a wealthy lumberman who invested in an indemnity just as the Sudbury fire was about to overrun his operation in the Temagami forest.[22] The 1936 forest fire season was significant not just for the amount of trees and prairie grass that burned, but for the 647,000 hectares of merchantable timber that was destroyed.[23] That's almost 100,000 hectares more than the size of Prince Edward Island.

Questions about who and what was causing some of these fires got pretty weird when the director of the Quebec Provincial Forest Fire Protection Service speculated that communists had started a fire in the Timiskaming area in retaliation for the prosecution of one of their members.[24] Stranger still was that this contention turned out to be true. Even weirder was the story of two criminals who escaped from the Burwash prison in northern Ontario, allegedly setting fires in the forest in an effort to prevent authorities from tracking them down. One of them was found dead in a lake, bruised and battered, with shoes and socks on but trousers missing. Fearing that his killer was "running wild" in the bush, intent on setting more fresh fires, local residents were in a panic,[25] which was understandable because there were already 5,000 men – the biggest firefighting force up until that time – fighting forest fires in northern Ontario.

The three fires in the thirties that had the biggest impact on public policy were the Boundary Fire in Glacier and Waterton Lakes parks on the Alberta/Montana border in 1935, the Bloedel Fire on Vancouver Island in 1938, and the Dance Fire that burned on Thanksgiving Day in northern Ontario later that year.

The Boundary Fire

In 1932, the governments of Canada and the United States linked Waterton Lakes and Glacier national parks to form the world's first international peace park. It was one of two peace parks created that year, signs of friendship and co-operation between two countries that had been duking it out over pulpwood tariffs before shared experiences in a world war and the Great Depression brought them closer together. The other was the International Peace Garden that was created in the Turtle Mountains of North Dakota and Manitoba. The irony in calling them "Peace Parks" is that both came into being only after Indigenous people were moved out of both areas under duress.

The first test of international co-operation came in the summer of 1935, when lightning triggered a fire on the Glacier side of the peace park. Chief Park Warden J.C. (Bo) Holroyd described how he and US ranger Lou Hanson talked on the phone before deciding to ride in on horseback and meet in the mountains along the border with small firefighting crews. It was not enough. As gale-force winds blowing in from the south fanned flames along the flank of Mount Richardson toward the town of Waterton, both men called for extra help. US Forest Service firefighters and fifty men enrolled in the Civilian Conservation Program were brought in by boat. More than a dozen men stationed at a Canadian relief camp in Lethbridge joined them.[26]

By the time rains came in early September, 895 hectares of forest had burned. No one got hurt, and limited damage was done to infrastructure. But the fire set the stage for an ongoing, ever-expanding relationship between Canada and the United States that eventually included Indigenous people from both sides of the border.

The Bloedel Fire

Vancouver Island was supposed to be too wet to burn big until it did, in 1922, when the Merville Fire killed a fifteen-year-old boy as it ripped through 8,500 hectares. The cost of suppressing that fire was more than $700,000 in 2021 dollars.[27] The Comox Logging and Railway Company claimed the fire had started seventeen kilometres north of the Oyster River in International Timber Company logging operations. The courts, however, didn't believe this assertion when the company was sued by Merville settlers. They won their compensation claims, alleging that sparks from a Comox Logging locomotive had started the blaze and that the company neglected to take proper precautions to contain and put out the fire.

The Bloedel Fire of 1938 was much bigger, more costly, far more destructive, and a further illustration of how recklessly the forest industry continued to harvest trees and how the liabilities associated with these practices were borne by the public. It also underscored just how poorly prepared the provinces were for dealing with wildfire without a Civilian Conservation Corps in Canada or a well-endowed national forest service. "Darkness at Noon" was how John Parminter of the BC Forest Service described it when he pulled together data on the fires that burned through the island for more than three weeks.[28]

It was the fifth of July when a Bloedel, Stewart & Welch logging company fire patrol discovered a small fire along the railway tracks running through the company's timber lease. The crew had no firefighting equipment on hand to deal with it. The only way they could send word of what was happening was by telephone, the nearest of which was located about three kilometres away at a logging camp. Patrolman David Crawley had to run there to relay the location. An employee at the camp then passed the message on to the Forest Service's branch at Campbell River. It was like a farmer driving into town to get the volunteer fire department to come and help him douse a fire burning a barn full of dry hay.

More than ninety minutes passed between the time the fire was discovered and when the first firefighters arrived. It took another two hours before a water pump was brought in. The fire was two hectares in size by that point and growing fast. There was still no sign of the bulldozer needed to create firebreaks. Firefighters had no chance when a powerful northwesterly blew in, sending embers across a lake that was a half-mile wide. Only then was C.C. Ternan, the assistant district forester, called and advised about how dire the situation had become. Ternan arrived the next day by float plane only to learn more bad news. The fire had imbedded itself in old slash piles that had been left behind by the forest company.

Lighting backfires would have been an option to prevent the fires from advancing. But for some reason that strategy wasn't used until the third day, when it looked like the town of Campbell River might be in harm's way. By day eleven, more than 400 men from Bloedel, Stewart & Welch were fighting the fires. Another one hundred men employed by the Forest Service arrived later that day. However, the wind kept blowing embers along, forcing the guests and owners of a hotel to evacuate to Campbell River – reluctantly, it seems. Many of the guests wanted to stay and watch the advancing fire. When the SS *Princess Elaine* arrived in Nanaimo with sixty more firefighters, chief forester Ernest Manning described the situation as "very seriously disturbing."

More help was needed, but there was nothing like the CCC to draw men from. In broadcasts from several Vancouver radio stations, the superintendent of the Employment Service of Canada begged for volunteers. Within fifteen minutes, 150 men signed up.

On the thirteenth day, the rapidly expanding blaze forced the evacuation of a dozen families from Forbes Landing. Manning opined that this could develop into "one of the worst fire seasons in history unless we get damper weather soon." By the beginning of the third week, there were 850 loggers and unemployed men fighting the fires. The smoke was so thick that a fish packer from Nanaimo ran aground near the Gabriola Island lighthouse. Highways were forced to close. Communications became increasingly difficult because so many telephone poles had burned down.

9.5 The Bloedel Fire wasn't the worst in Island history, but it was one of the most memorable. Winds of 60 km per hour fanned the conflagration and 1,700 firefighters battled it.

The Canadian destroyers HMCS *St. Laurent* and HMCS *Fraser* were summoned by the Forest Service to assist with fire suppression and an emergency evacuation of more residents and firefighters. The plan was to move the ships to Campbell River and put hose lines ashore to safeguard the town. Police were being called in to help when the fire expanded to 20,000 hectares. And then the unthinkable happened. Someone slashed 1,000 feet of hose at one of the fire camps and poured sugar into one of the water pumps. Furious, fire boss Charlie Haddon dismissed approximately one hundred of the volunteers. Many pleaded for him to reconsider, offering to work for free to clear their names. "Some of you boys haven't been hitting the ball, and you're going back to Vancouver," responded Haddon. "I'll be frank with you. Some of you may not be at fault, but the situation is too big and too dangerous to take chances."

It was not the most prudent of decisions. Every bit of help was needed. By this time, the fires were threatening old growth timber, which would burn much more intensely than the relatively young and more resilient second-growth forest. In the days that followed, high temperatures and strong winds persisted,

9.6 The Bloedel Fire enveloped almost two-thirds of Vancouver Island and was reported as far as 640 km south in Portland, Oregon. Untold numbers of animals died, but many were saved.

forcing more and more people to evacuate. Echoing Manning's earlier prediction, Premier Thomas (Duff) Pattullo described it as the worst fire since white people started settling in British Columbia. He called for public co-operation.[29] But, inexplicably, logging operations continued and campfire permits were still being issued. It wasn't until day seventeen that Pattullo finally ordered a stop to both. Even then, he seemed reluctant. Commerce was apparently more important than public safety. "The timber industry will always be one of our chief industries and we cannot afford to jeopardize an industry which presently employs some 30,000 individuals, produces $75 million a year, and is capable of perpetuation," he said, short of apologizing for doing what seemed to be obvious.

The day Pattullo announced the decision to the public, the fire grew by a third. "Black Friday" was how the firefighters described it, and it had, by this time, become world news. A correspondent for the London *Times* was not so much concerned about the welfare of people as they were about the fate of the world-class trout fishery at Forbes Landing.[30] The *New York Times* and *Los Angeles Times* reported on it for several days, noting how similarly large fires in

Washington and California had been brought under control quickly with the help of fully trained Civilian Conservation Corps personnel.

By this time, the Corps was seen as indispensable in fire prevention and fire suppression. Prior to the 1938 fire season, in southern California alone they had constructed and maintained 183 kilometres of fire breaks, two million litres of water capacity, eight fire tower lookouts, and 125 kilometres of fire truck trails. They were also maintaining 340 kilometres of existing trails in addition to fighting fires.[31]

There was nothing in Canada to compare. British Columbia had counted on its wet weather to protect the west coast from big fires, but the rain didn't arrive on time that year. When it was all over, chief forester Manning refused to blame the logging companies. "Let us not be too critical of this high-lead logging method with it is accompanying wastefulness until a substitute is found and not compel them to close down, thereby adding to our unemployment problem."[32]

The Dance Fire

The day after fire tore through Dance Township in northern Ontario on Thanksgiving weekend of 1938, two Fort Frances police constables arrived at the farms of brothers William, Frank, and Noah LaBelle and found eleven badly burned bodies scattered across the road. "The two mothers had tried to protect their babies by covering them with their bodies," one of the constables said. "It was a terrible sight. Every few feet or so we would find a body twisted and burned either in the ditch or on the road. It was a burnt road of death."

As destructive as the Bloedel Fire was, the Dance Fire was deadlier. By the time rain extinguished the fires, twenty people, including three from the American side of the border, had died in situations that were entirely avoidable. Lessons that should have been learned from the Haileybury Fire of 1922, the Matheson Fire of 1916, the Porcupine Fire of 1911, and the Baudette–Rainy River fire that killed forty-three people in the same area in 1910 had largely gone unheeded. Once again, the disaster was caused by farmers burning slash piles from the remains of timber they had cut to get extra cash from the local pulp mill. Because it had been so hot and dry that summer and fall, rangers advised people to get out when it looked like the fires were going to run toward the towns of Rainy River and Emos. No one in the countryside, however, was willing to leave their livestock unattended. Some of the farmers did their best to stop the slash fires from spreading, only to have several days of effort lost when gale-force winds suddenly blew in.

Dan Patterson was working as a woodcutter for one of the LaBelle families. In an interview with the *Fort Frances Times and Rainy Lake Herald*, Patterson described how just hours before the fires came, they had had their noonday meal and were about to go back to work cutting pulpwood. "It was a beautiful afternoon, and sort of oppressive," recalled Patterson who confessed that he had an uncomfortable "premonition that something was going to happen." Slash fires had been smouldering for a month. As he and a co-worker took a shortcut through the woods to get to the meadow they were working in, they could hear "something desperate, like the roar of a dozen freight trains going over trestles in the distance." Patterson ran back to the farm to warn the others. While he was hitching up a horse and wagon, Frank LaBelle gathered the women and children from three local families and started walking down the gravel road to get away from the advancing fire. Shortly after Patterson caught up with them, the fire crossed the road and forced them to retreat.

They took refuge in a clearing, hoping that it was big enough to protect them. It wasn't, not by a long shot, given how hard the wind was blowing. When the house and buildings caught fire, "everyone, women and children, seemed to go temporarily mad, and they scattered in every direction," said Patterson. "Man, you'll never know what a forest fire is unless you've been in one. It is so terrifying that there is no possible way to describe it." Patterson was on the roof pouring water on the house when an intense wave of heat knocked him down to the ground, where he had the wherewithal to douse himself with water before falling unconscious. When he came to, he ran for help.

That was the last he saw of most members of the LaBelle families.

Four members of another family, and another single man, suffered a similar fate. Six more would have also perished had a lawyer returning from a hunting trip not picked them up in his car. When fire blocked the way out, he abandoned the vehicle, saving them by taking refuge in a clearing that had been burned by a previous fire. An editorial in the *Globe and Mail* suggested the disaster was avoidable, given the warnings and the history of devastating fires in the region. "A dark and discouraging picture," is what the editorial concluded.[33]

The optics did not look particularly good when Prime Minister Mackenzie King and O.D. Skelton, the Canadian undersecretary of state for External Affairs, set off from New York to vacation in Bermuda two days after the bodies were found, and just a day after yet another fire was threatening to destroy the village of Graham, 240 kilometres northwest of Dance Township. King was telling newspaper reporters how "anxious and arduous" his days were before his departure, and how much he was looking forward to a vacation. It was only then that he learned from reporters about the fire that killed so many people.[34]

Mitchell Hepburn, the newly elected populist premier of Ontario, looked even worse when the *Globe and Mail* reported that twenty forestry-school graduates working for the government had been let go the year before. That included the entire team investigating forest maintenance. The government had also removed some regions such as Rainy River from fire protection, allegedly for cost-cutting reasons. These blockbuster revelations came later from John C.W. Irwin, the secretary of the Forest League society and a member of the Saskatchewan Royal Commission on Forestry. He was a past member of the general executive of the Canadian Society of Forest Engineers and a past chairman of the southern Ontario section. He was also speaking for the graduate students who lost their jobs. "These investigators [laid off from the Ontario government] were the only men who could, properly speaking, be regarded as practicing scientific forestry in the north," Irwin told the *Globe and Mail*. "Fire protection and the administering of timber tracts as they are administered is not scientific forestry, and I defy anyone to find a definition indicating that they are more than parts of a problem."[35]

Irwin was emphatic in stating that none of the forestry graduates would deny what he was saying "unless intimidated by fear of losing his job." He went so far as to suggest that Minister of Lands and Forests Peter Heenan had come to the conclusion that forest fires were "almost unavoidable," and that there was, therefore, no use trying to control them.[36] He was right. Politicians in provinces like Ontario, Quebec, Alberta, and BC had often expressed an interest in scientific forestry. But just as Ernest Finlayson had predicted when management of western forests was handed over to the provinces in 1930, they were not interested in hiring scientists.

"The general approach by governments has been to spend as little money as possible in advance, but when fire comes, to parade an all-out determination to fight for the people's interest, regardless of cost," Irwin said, foreshadowing what would become routine practice across the country. "In all aspects under which fire protection is usually considered – prevention, detection, and communication, all the provinces, have lagged behind a reasonable minimum."[37] Even the financial editor of the *Globe and Mail* was with Irwin on this point. "Why bother with conservation, disposal of slash, protection against fire, when there seemed to be illimitable lands beyond for exploitation as one after another was cut over," Wellington Jeffers wrote.[38]

Mackenzie King apparently had enough of the bad publicity. In 1939, his government finally launched the National Forestry Program which Finlayson had so much wanted. The program hired 5,000 men on relief to work in federal and programs dedicated to forest protection and reforestation. Although the program ran for only a year because of the Second World War, the federal government, it seemed, was finally ready to do something significant about fire. It was the provinces that remained the problem. Most were slow on the uptake when it came to taking advantage of what the Canadian Forest Service had to offer. By 1939, when the National Forestry Program in Canada was created, only three provinces were using the fire danger index tool that Wright and Beall had invented.

10

THE FALL OF THE DOMINION FOREST SERVICE

On 26 February 1936, Ernest Finlayson left his home in Ottawa to go to work. He never made it to the offices of the Dominion Forest Service. As day turned to night and there was no word from him, police were called, and a search of the district was made over the course of the next week. "That he managed to drop so completely out of sight had both police and friends baffled," the *Ottawa Citizen* reported.[1] No one could have predicted this from the director of the Dominion Forest Service, who had reached that position in 1924 after rising in the ranks for more than a decade.

Finlayson had a Pinchot-like look to him. He was handsome, smart, and mustachioed, as Pinchot was. The first thing he did as director was to host a forest protection conference in 1924, where he suggested that the federal government might be willing to extend firefighting aid to the provinces. But his desire to transform the Dominion Forest Service into the national agency that Pinchot and his successors had craftily built was not meant to be in a penny-pinching government that did not want to pick a fight with provinces over control of natural resources. And the fact of the matter was, neither government nor industry was interested in investing in the present for the long-term health of forests.

In 1929, the Dominion Forest Service had a budget of $1.5 million, which was a fraction of what the US Forest Service was working with. A year later, when responsibility for federal forest reserves was transferred to the western provinces, the amount dropped to $325,000.[2] The Dominion Forest Service was no longer responsible for the administration and protection of forest reserves, nor for fire

10.1 Ernest Finlayson, director of the Dominion Forest Service from 1924 to 1936.

protection or forest reconnaissance. Its operation of nurseries was transferred to the Department of Agriculture.

Finlayson fumed, calling the decision to hand over forest resource management to the provinces a "retrograde step." He tried to keep the agency on track. But the situation got worse as it became clear that the provinces wanted money, not advice, from the federal government. It got to be so bad that many of the records of the Dominion Forest Service went to the local dump when federal offices closed across the country.[3] The provinces had no interest in taking on archival responsibility and no other federal agency was willing to come to the rescue. It was left to the newspapers to maintain a record of what had happened in the past.

Indeed, up until 1934, the only reliable allies Finlayson had were the newspapers, Frank Barnjum (who he didn't trust right up to the time he died in 1933), and William Finlayson, the Ontario minister of Lands and Forests (who was his half-brother). The family support did not last long. William was defeated that year and Peter Heenan was elected and appointed to his position. Heenan was a man who John Irwin later described as someone not caring a wit about wildfire.[4] A stalwart in the Liberal government of Mitchell Hepburn, he was determined to cut the forestry budget to the bone. As historians Peter Gillis and Thomas Roach noted, even a "university degree was not seen by him as a prerequisite to efficient firefighting."[5] Just like the premier, Heenan was more interested in awarding timber contracts to government friends than investing in forest science and management. Allegedly, Hepburn became a wealthy man while in office because in return for favours, he got stock market tips.[6]

Finlayson realized there would no McSweeney-McNary Forest Research Act (1928) in Canada to set the stage for government-sponsored wildfire science and a rich network of forest research stations such as the US Forest Service had been building since 1908. Nor would there be, save for one year in 1939, a Civilian Conservation Corps–like program in Canada that would put unemployed men to work planting trees and suppressing fires. All that the Dominion Forest Service would get during the latter part of the Depression was a paltry amount of labour for the establishment of extremely modest experimental research stations at Valcartier, Quebec, the Duck Mountain Forest Reserve in Manitoba, Acadia in New Brunswick, and Kananaskis in southwestern Alberta.

As morale in what remained of the Dominion Forest Service sank in the 1930s, Finlayson became afflicted with ulcers, amnesia, and a nervous condition that hinted at an impending breakdown. He must have realized that survival was all that he could hope for. The only way that could happen was with breakthrough research. This path had already opened up in 1917, when military officials approached the Dominion Forest Service to undertake forest fire control and management of the wooded area around the Canadian Forces base at Petawawa, outside Ottawa. When the station was established in 1918, it was little more than a tent camp with a budget of just $6,000 to conduct research. And because of the inclement weather and the lack of proper accommodations, research was limited to late spring, summer, and fall.

The Petawawa station, however, was gifted with talented men. James Wright, a civil engineer with the Dominion Forest Service, was one of them. Wright's interest was in how and in what ways weather favoured fire. In a Christmas Eve note to Finlayson in 1925, Wright planted the seed for an operational fire research program within the station. Finlayson, who by then saw forest fires as a national problem, gave him the green light. This small investment in science eventually paid big dividends. Together with his protégé Herbert Beall, a University of Toronto forestry school graduate, Wright created a series of increasingly sophisticated fire-danger rating systems after figuring out how weather determined moisture content in trees, shrubs, grass litter, and duff (dead plant material) and how that led to the ignition and spread of fires over the Canadian forest landscape.

This breakthrough and others were significant enough to get the attention of the *New York Times* in 1935. "Hereafter, forest supervisors in Canadian woods will go armed with slide rule, charts and tables. These will be used in estimating from day to say the degree of fire hazard in certain areas."[7] This was important because politicians paid attention to what newspapers such as the *New York Times* had to say. Whether the article had anything to do with Canada's National Research Council holding a conference on forestry research later that fall is unclear. But the newspaper coverage did not hurt.

In the years after it was established in 1916, the National Research Council picked up where the Commission of Conservation left off before it was abolished in 1921. The 1935 meeting was significant in that it had the support of both the Canadian Pulp and Paper Association and the Canadian Society of Forest Engineers. It was attended by most of the leading lights in forestry, and it ended with the NRC's establishment of a subcommittee on forest protection.

Fire historian Stephen J. Pyne suggests that the subcommittee "furnished a neutral setting, independent of political authorities, where all could speak and no one had to listen." For the next fifty years, this group and its avatars provided the only collective council for Canadian fire, the closest approximation to a national organ. Here, rather than through the Dominion Forest Service, the Canadian Fire Community could describe their year's activities, ambitions, and alarms.[8]

True enough. But it didn't matter much what they had to say during the Depression years, when provincial governments like those in Ontario were slashing budgets and letting pulp and paper and mining companies do as they pleased. If there was a national organ on wildfire at this time, it was not the subcommittee of the National Research Council; it was the media that continued to speak directly to the public and politicians. Since the Haileybury Fire of 1922, newspapers had doggedly chased the wildfire story. The Depression did not dampen their enthusiasm. With more and more people working, living, and recreating in forest environments, wildfire continued to wreak havoc. It was happening too often to ignore. The Bloedel, the Dance, and the many other fires that burned in the 1930s were proof of that.

The budget for the National Forestry Program, which Finlayson so much wanted, was a far cry from the estimated $54 million that would have been needed annually if Canada had attempted to replicate what the United States had achieved with the Civilian Conservation Corps.[9] It would have been a start, had the end of the program not been so close in coming.

The program ran for only a year because of the outbreak of war. Although it never had a chance to do what it was intended to do, it demonstrated that the federal government was finally on board when it came to improving forest management. It was the provinces that remained the problem.

11

THE ROYAL COMMISSION INTO WILDFIRE

In the early 1940s, the fire prevention and fire suppression system across Canada was a mess. Quebec's École forestière de La Tuque was the only forest ranger school in the country. Apart from its graduates, other rangers were mostly seasonal recruits who continued to be poorly paid, untrained, and ready to move on to new jobs that offered more secure futures. Too much emphasis was being devoted to cost-cutting. A low point came when British Columbia limited the number of pencils allotted to forestry personnel to two. In 1947, the province of Ontario was still not using Wright and Beall's fire index. It had only two-thirds of the lookout towers necessary to detect forest fires. Remote telephone communication was still lacking in much of the province.[1] Air patrol detection was down to about 5.7 hours per day to patrol 160,00 square miles of forest.[2]

These and other shortcomings were emphasized in 1945 by Gordon McGregor Sloan, a judge and former member of the BC legislature who led a royal commission on the Forest Resources of British Columbia into the province's forestry practices, and by John Irwin, a non-practising forester, co-founder of the Clark-Irwin Publishing Company, secretary and editor of the Save Ontario Forest League.

Sloan's 10,000-page report was supposed to transform the BC Forest Service from "a half-starved revenue collection agency into the kind of an organization all foresters feel it should be."[3] The BC government, however, ignored his recommendation for a forestry commission that would manage the forests without political interference. Instead, it proposed a new system of forest management licences that ended up being dominated by corporate interests at the expense of government oversight. Even then, companies were not happy, as Oscar Lundell, a lawyer for British Columbia Forest Products, recalled years later. "Companies were a little leery of it at first and they still are," he said of the Forest Act that came into play in 1947. "The Crown is landlord and can fix stumpage,"[4] the price on standing timber and the right to harvest it.

Irwin was much more focused on fire prevention and suppression when he spoke to the *Globe and Mail* in October 1945. Irwin told the paper that the amount of forest burned in Ontario in the previous thirty-five years had covered one hundred times the area that was replanted. Twice as many lookout towers and twice as many telephone lines were needed to "give adequate coverage but by no means generous coverage to our organized forest area of 170,000 square miles."[5] The *Globe and Mail* evidently viewed all of this as credible. The paper published seven of Irwin's articles over a week-long period to demonstrate to the public and politicians how bad the situation was.

Conservative premier George Drew, a former soldier, was not one to sidestep issues, as Prime Minister Mackenzie King so often did. But this was an exception. Like many premiers and prime ministers who needed to buy some time, he called for a royal commission on the subject of forestry management. Unlike the royal commission of 1924, which was soft rather than hard on recommending changes, the royal commission of 1946–47 was remarkably comprehensive. Major General Howard Kennedy, who oversaw it, seemed initially to be a run-of-the mill choice to lead the commission. He had been surveyor for the E.B. Eddy Company before becoming its woods manager in 1926 and held that position until war broke out in 1939. But Kennedy turned out to be a surprise. Rather than holding the hearings solely in metropolitan centres such as Toronto or Ottawa, he had the commission team convert a Canadian National Railways colonist car into a mobile office. At each stop along the way from Quebec to the Manitoba border, and from southern Ontario to Ottawa, they were greeted by a district forester who had set up a tent for them to dine in. Stops varied from eight to fourteen days at thirteen different locations. Where the railroad didn't go, they travelled by car, plane, and boat as far south as the American border and as far north as Fort Severn on the coast of James Bay. A dozen visits to extremely remote locations were made by float plane.

The commissioners inspected new and old cuttings and evaluated labour conditions, housing, and firefighting resources. Every pulp mill and every sawmill of significant size was inspected. When not in the bush, the commission held public hearings in Sault Ste Marie, Port Arthur, Kenora, Fort Frances, Geraldton, London, Cochrane, North Bay, Pembroke, Ottawa, and Toronto. Notes taken during the day were transcribed into formal recordings at night. Neither Sundays nor holidays stopped them and more than a half-million words were recorded. Kennedy was not immodest in suggesting that this was "probably the most complete and up-to-date data that has ever been assembled" on forestry.

Kennedy's report vindicated every charge that Irwin had aimed at the government. It highlighted the shortcomings of forest fire prevention and suppression not just in Ontario, but in all of Canada. The cadre of forest firefighters, according to Kennedy, was understaffed, poorly paid, and mostly seasonal. Many of the men were in constant fear of losing their jobs. The training was woefully inadequate, and standards for employment were so low that too much money, according to the report, was being spent on employing "misfits."

In addition, the line of command between regional and district foresters was unclear. Access to fires was limited by the number of roads going into a forest. Radio contact both on the ground and in the air was inadequate. The province still wasn't doing what the US did routinely: parachuting equipment into areas where there were no roads. Furthermore, equipment upgrading was badly needed. There were not enough bulldozers available to build fireguards. Insufficient effort was expended in plotting all fires and their causes to determine whether some transitory or non-recurring factors had been given too much importance. Kennedy emphasized that fuel studies to better determine fire danger were lacking.

The one outlier was a curious recommendation calling for the systematic killing of porcupines, which the commission estimated were destroying tens of thousands of trees. Still, given the scope of the inquiry, the common-sense recommendations, the integrity of Major General Kennedy, and his close connections to both the forest industry and the government, some good should have come from the commission. A forest ranger school that John Irwin had lobbied for was already in the works in Dorset, a three-hour drive north of Toronto. But in a system that had been broken for so long, most of the fixes recommended by Kennedy's commission were apparently too costly to be accepted by big companies that had always done what they pleased in the forests of Canada. Like other provincial governments, the one in Ontario was unwilling to take them on.

John. D. Gilmour, a retired forestry engineer who had been hired by the BC-based MacMillan and Bloedel company to develop a forest policy, said as much in 1950 when he wrote an article in the *Financial Post* with the headline: "Facing Ruin through World's Worst Forest Laws?" In the article, Gilmour echoed what Frank Barnjum had been complaining about in the 1930s and what Irwin was saying in the late 1930s and 1940s. "The whole of Canada is notorious for poor forest laws and poor management. Among foresters this is well known, including the fact that Ontario is the worst of them all. This means that Ontario has the worst forest policies and administration of all the civilized regions of the world. Disaster, therefore, must be inevitable, and signs of it are clear to all foresters who know the situation."[6]

Douglas White Ambridge, the president and general manager of Abitibi Power & Paper Company Limited, wasn't going to sit back and take this and other barbs being thrown at the industry without a fight. He had been in the forestry business too long to admit that things needed to change. Following three years of service during World War I, he graduated from McGill University and got a job with the company he would eventually run after stints with several other resource and power companies. The issue for him was not the conservation of forests; it was that there were not enough trained men to double the harvest of trees. In a speech to the Empire of Club of Canada, Ambridge mocked Gilmour for breaking rank with a tightly knit cadre of forest company executives who would not admit to any wrongdoing. "Please note the inevitable superlative … 'Ontario has not got the second worst, it has the very, very worst forest laws,'" said Ambridge. "Now, Mr. Gilmour is an old friend of mine and he is not an alarmist over the ordinary goings-on in life, but when it comes to the forests he throws caution to the wind, unlimbers his superlatives and lets fly. His forbears, it seems to me, may have been Druid priests who knew well how to enforce their religious disciplines by scaring the hell out of their congregations."

He even invoked the name of Frank Barnjum, belittling him for his conservation views and suggesting that he came from a long line of men "whose ancestor was probably a king of the wood in Roman times." Ambridge did not stop there. He dismissed the royal commission's report as "unnecessary and untrue" and "so impracticable that nothing more has ever been heard of it. I do not believe that General Kennedy himself has ever publicly advocated the adoption of his so-called plan since the report was published … I have no time to describe to you what this plan was except to say that it was a 'dilly,' but when I hear complaints that these very expensive Royal Commission reports are almost always pigeon-holed, I ask myself what else could be done with the 'Solution' proposed by the last Royal Commissioner to sit in judgment on Ontario's forests."

And then he told the men in the room everything they wanted to hear, and what executives in the forest and energy industries would continue to say in the decades that followed, even though they certainly knew it was not true. "We should realize that the men who are charged with the responsibility for managing the pulp and paper industry in Canada are not a lot of bloated capitalists complete with silk hats and long cigars. By and large the managers of the Canadian pulp and paper industry are men who started at the bottom and worked their way up. And when I say worked, I mean worked. They are not all men of wealth but all of them without exception are good citizens of Canada who are just as interested in maintaining the health and productiveness of the forests upon which all our livings depend, as anyone else is."

The problem in the forests of North America was not industry, he said; it was fear and loathing and public ignorance. "As one of the group of senior executives of the greatest Canadian industry I have felt that we have not done a very good job in supplying the public with facts as an antidote for their fears and fancies. This same thing is true in the United States, and in recent years the pulp and paper industry down there has been paying a lot more attention to publicizing the facts of the forest and has made some progress in removing the fears and fancies, regarding the forests, of the average American citizen."[7]

12

WHITE MAN'S FIRE

In August 1937, Charles R. Lathrop, the assistant forest fire warden for the state of Connecticut, said something no one had ever said before in public. "Women make better (fire) lookouts than men." Lathrop shared that sentiment in a lecture on the state's woodlands. "They are more conscientious and more contented with the 9 a.m. to 6 p.m. hours of duty."[1] Eleven of the state's nineteen lookout towers were "manned" by women. Connecticut was ahead of its time and, with few exceptions, very much out of step with most of North America.

Forestry was not only a man's world; it was dominated by white men who blamed careless women and Indigenous people for lighting fires. People of colour were largely excluded from participating in fire suppression even after the Black Twenty-Fifth Infantry famously evacuated the town of Avery during the Big Burn of 1910. The one notable exception was Charles Young, a West Point graduate of distinction who in 1903 became the first African American to serve as superintendent of a national park. The job in Sequoia National Park was offered after he and his troops were brought in to patrol and construct roads and trails. His hiring of and consulting local townspeople endeared him so much that they proposed naming one of the giant trees in the park after him. Young declined, insisting that the tree be named after Booker T. Washington, a former slave who advised President Theodore Roosevelt on racial issues.

Roosevelt wasn't as progressively opened-minded as all that. "Race suicide" was a term he used to describe women who chose not to bear children. Gifford Pinchot's views on people of colour were evident in his participation in the first

12.1 Helen Dowe at the Devil's Head Fire Lookout, Pike National Forest, Colorado, 1919.

and second International Genetics Congress and as a member of the advisory boards of the American Eugenics society. Henry Greeley was receptive when he met with a group of female US Forest Service workers who wanted him to consider expanding their responsibilities, but he made it clear that they "could never expect to see a woman filling the position of chief fire inspector on the National Forests." Instead, he said, "I expect to see women running public relations one day."

But Connecticut was not the first state to hire women to manage lookout towers. In 1913, Hallie Morse Daggett was hired by the US Forest Service to command the Eddy Gulch Lookout Station at the top of Klamath Peak in Klamath National Forest in California. It was not an easy place to get to: three hours in on foot to reach the top of the 6,444-foot mountain. In spite of predictions that she would fail, or get too lonely to tough it out, she did extremely well even when telephone lines went down and cut off all communication for days. She returned for twelve more summers before she retired from the job.

In 1937 Ontario, not one of the 1,000 fire rangers was a woman. In fact, such a thing was never contemplated, stated a government official when asked for a comment on the role that women might play in fire prevention and suppression. The question was greeted with a "ripple of laughter." "Fire rangers in Ontario have many duties to perform other than 'look-out,' which necessitates a man for the job," according to the unnamed government official. "In between times they are engaged in trail-cutting, construction work, putting up telephone lines, and as for actual firefighting, only a man could do that."[2]

A story soundly rebutting that point of view came many years later, when firefighters in northern Ontario were fully taxed. At the time, the wife of the deputy forest ranger was alone at the family cabin with a four-year-old and a ten-month-old baby when she spotted smoke rising ominously from across the lake. She phoned the local fire tower, loaded her children into a canoe, and paddled across the lake. She then started pouring water onto a fire burning in between a cabin and the lakeshore. By the time help arrived, she had brought under control a small fire that would have otherwise turned into big one.[3] The forest commissioner commended her for preventing a possible disaster. But he and the newspapers never referred to by her first name. To them, she was Mrs Waino Autio, wife and mother, and nothing more.

Women were held in such low esteem in the world of forestry that in 1936, the state of Michigan blamed vacationing female smokers for causing so many fires. There was no evidence for this, and it's hard to imagine it could have been true. But according to a report in the *New York Times*, "fire prevention authorities feel that a brisk education campaign among women must be conducted."[4]

Forestry was not only a man's world, but it was also world dominated by white men who blamed Indigenous people for lighting fires that caused destruction. In the midst of the fires that burned so much of Canada in 1911, the Dominion Forestry Branch let it be known to newspapers that it was launching a campaign to educate Indigenous people about the "the disadvantages to the natives that follow the burning of the forest." Plans were being put in place, the newspapers were told, to enlist the Natives as volunteer fire rangers. "Handsome badges," not a salary, were going to be issued to entice them to sign up.[5]

African Americans suffered similar indignities. In the 1930s, the Civilian Conservation Corps hired 200,000 African American men, but CCC director Robert Fechner, a conservative southern labour leader, insisted on segregating them in their own camps. Policy dictated that they would not be "forced upon" local communities that did not want them. Nor were African Americans hired according to need, even though they were often in much more dire economic circumstances than their white counterparts.[6]

This attitude was evident when African Americans from the 555th Parachute Infantry Battalion were dispatched to deal with Japanese fire balloons on the Northwest coast. "Both officers and enlisted men were making bets that we wouldn't jump – we'd be too afraid," recalled Clarence Beavers, the last surviving member of this ground-breaking group.[7] Racism was clearly behind the decision to send Beavers and his colleagues out west. "Major commanders in

12.2 African American Civilian Conservation Corps Fire Fighters, northern Minnesota, ca 1933.

Europe were leery of having highly trained colored paratroopers coming into contact with racist white elements of the time," according to the 555th Parachute Infantry Battalion Association's website.

The response to wildfires in North America was and still is cultural. Indigenous people in Canada represent only 4 per cent of the population. But 40 per cent of the evacuations that result from wildfire involve Indigenous people. Up until 2021, the Canadian government considered it easier to exile these people to cheap hotels down south for weeks, in some cases separating family members from one another, rather than making their communities more resilient to fire. The situation has improved, thanks in large part to the Aboriginal Firefighters Association of Canada (AFAC), which was founded on 19 September 1991 in Portage La Prairie, Manitoba.

They have made great strides. But in a survey of aboriginal firefighters conducted in 2021, nearly half indicated that they experienced work-place discrimination which included mocking, bullying, and being subjected to humiliating jokes. Some refused to report the abuse out of fear of retaliation or because they didn't think that authorities would deal with it. Even then, most were "satisfied or extremely satisfied" with the work because it was challenging and they liked to help people and communities in need. They also felt a strong responsibility to care for Mother Earth.[8]

More and more women have proven that they can manage and fight fires just as well as men. But they still represent less than a fifth of the firefighting community in the Bureau of Land Management that is responsible for a tenth of the landmass in the United States. Few rise to upper management levels, and those that do, like Kelly Martin, Yosemite chief of Fire and Aviation, have complained of sexual harassment, intimidation, and suggestions that they got the job only because they were female.[9]

Howard Hedrick, second in command of the Bureau of Land Management (BLM) fire program, argued that experience had taught him "that if you have the same kind of people with the same backgrounds, experiences and education, when faced with challenges, they're most likely going to come up with some of the same solutions. If you have a more diverse group, I think you'll come up with much better solutions."[10] The forest management world did not have that diversity through most of the twentieth century. Better solutions came, but they were slow in coming. Everyone preferred business as usual to engaging in new ideas and embracing radical change.

13

INTERNATIONAL CO-OPERATION

On 25 October 1947, the Norwegian tanker *Solstad* was approximately one hundred kilometres off the coast of Maine when the crew experienced the distinct smell of a forest fire. The sailors were perplexed but learned that they were not mistaken when they radioed in to find out what was happening. They got news that most of the forested splendour of Acadia National Park and much of the seaside town of Bar Harbor in Maine were gone. A fire had swept down the slopes of Cadillac Mountain into the internationally famous resort town, burning through everything in its path with astonishing speed.

A Dunkirk-like evacuation had taken place two days earlier, when fishing boats and pleasure crafts of all kinds came to rescue more than 2,000 people huddled on the docks downwind of gales sweeping the flames towards them. As the boats began transporting some of the evacuees across Frenchman Bay towards Gouldsboro on the eastern mainland, Coast Guard vessels, Navy aircraft, and a Navy destroyer were on route from as far away as Boston and Rhode Island to help hundreds who had fled the area in 700 trucks, buses, cars, and bicycles. Bulldozers were brought in to push away the burning timber that blocked their path.

It wasn't just Acadia National Park and Bar Harbor that were in the line of fire that year. Much of northeastern North America was ablaze or smouldering. In August, 225 people had been evacuated from the village of Pleasant Bay in Nova Scotia as fishing boats and the RCMP cutter *French* patrolled the waters of three Gulf of St Lawrence communities, ready to evacuate more people if necessary. Another one hundred people from the Black Loyalist community of Tracadie had been forced to flee just days earlier. In the Newfoundland village of Bonavista, every "able-bodied man" was called upon to help stop a fire from entering the community. Firefighters along a highway in northern Ontario were so overwhelmed that the mayor of Hearst asked permission from Ottawa to enforce conscription of every male inhabitant of the lumber town. Locals were rounded up, loaded onto trucks, and sent to fight the fires.[1]

The state of Maine, however, was the centre of attention that year. In October, there were 200 fires burning hundreds of homes and businesses as well as the city of Portland's wooden pier. At the height of the conflagration, 2,700 soldiers were working shoulder-to-shoulder with the state's fire crews, National Guardsmen, and civilian volunteers. All eastern Air Force fields, including Washington, DC's Bolling Field, were ordered to be on alert. Military planes transported thousands of feet of pump hose from Wisconsin. As the situation became increasingly dangerous, Governor Horace Hildreth went on the radio and urged people to remain calm but to prepare for disaster. He called on residents to mobilize their wartime defense forces to "fight the greatest human and economic tragedy that has ever faced Maine."

"I'd been overseas, and I think I was as scared during the fire as any time when I was over there, at times," said a volunteer by the name of John Smith. "You just figured that you weren't going to get out of it. Because you figured there was nothing that was ever going to put this fire out, you kinda were getting the feeling the whole state was going to burn. In fact, there wasn't much that stopped it until it got to the ocean."[2]

Shortly after Hildreth made his plea, President Harry Truman declared all of Maine a disaster area, as the loss of property throughout the state and other parts of New England was well on its way to $27 million. By the time the fires stopped smouldering in mid-November, more than 81,000 hectares of forest, 851 homes, and 397 cottages had burned. More than 2,500 people were left homeless. But remarkably few people died. One elderly man perished when he returned home intent on saving his cat and an air force officer and a local teenage girl died in a car accident. A man and woman with underlying health problems succumbed to heart attacks. Several soldiers were injured when the truck that was transporting them crashed.

In all, 279,000 hectares burned across the northeastern US and eastern Canada within a matter of days. It came to be known as "The Week Maine Burned."

There was not a lot to differentiate that fire season in Maine from all the others that had caused so much damage across North America. The premonition of disaster began, as was often the case, with an intense drought that followed a very wet spring. Everything was so bone dry by September that Kennebunkport Fire Warden Arthur Welch was telling everyone, including the local police,

1947 Bar Harbor Fire Anniversary Supplement

The Bar Harbor Times

BAR HARBOR TIMES, THURSDAY, OCTOBER 19, 1967

ENTIRE ISLAND MENACED BY FIRE

AERIAL VIEW — Loaned by S. Wesley Brown, J

Thousands of Acres of Timberland Burned: Hulls Cove Threatened: Fire Still Raging

October 23- Bar Harbor Times

What is the most spectacular forest fire, and possibly the most dangerous that Mount Desert Island has ever experienced, threatened Village of Bar Harbor Tuesday [illegible] after traveling from across [illegible] highway between Beaver Dam, [illegible] and Dolliver's Dump through to the Eagle Lake Road, jumping the Breakneck Drive and straddling the Norway Drive and then turning west with the wind, burned to the edge of Eagle Lake on the south and the New Eagle Lake Road on the east, where the Bar Harbor Fire Department, assisted by volunteers and apparatus from neighboring towns as far away as Bangor and Brewer, checked it at dawn, Wednesday morning and temporarily ended its threat to advance into the westerly portion of the village proper. The timber, dry brush and forest areas permitting fire to rush uncontrolled [illegible] a disastrously strong sud[illegible] from the Northwest.

...As the [illegible] efforts are being [illegible] to press all [illegible] on Hulls Cove which at [illegible]

Fire Department, assisted by Captain Cassidy of the Bangor Fire Department, directed the hundreds of fire fighters with their apparatus, furnished by Northeast Harbor, Southwest Harbor, Somesville, Seal Harbor, Otter Creek, Ellsworth, Bucksport, Camden, several large detachments from Dow Field, as well as civilian volunteers. The fire which had burned for two days, was apparently controlled in the Great Marsh on the road from Hulls Cove to Burn's Corner, turned to its intensity by the wind—which suddenly started to blow shortly after daybreak Tuesday morning, jumped to the south of the road and raced southeasterly to threaten the farms along The Norway Drive belonging to Paul Russell, H. C. Whitney, Daniel McKenzie, and Howard Russell. At the latter point a barn and dairy operated by Hugh E. Kelly at a considerable distance from the main fire suddenly burst into flames, having been ignited either from the intense heat or a flying ember, and burned in a matter of minutes. Lost in the path of the fire was a small house on the Norway Drive owned by R. E. Wilcomb and the Thompson homestead, both of which were unoccupied.

In spite of backfires and desperate stands made by fire fighters, the fire jumped from tree-top to tree-top across the narrow points forcing the men and apparatus to withdraw and make new stands at more southerly points.

At about 5 o'clock Tuesday night the fire on a three-mile front approached the Hobe Young place and Malcolm Salisbury's farm on outer Eagle Lake Road, which were protected by firemen with booster pumps and by the scrubbing out of a fire by a bulldozer. At this point the fire veered more to the west and drove the fire across Eagle Lake Road and east to the property of Pearl McFarland and the Eagle Lake Riding Stable, situated on "McFarland's Hill." Again through the concerted efforts of several fire companies the buildings were saved, even though the fire surrounded them and jumped the road toward Aunt Betty's Pond at several points. Attempts were made there to halt the fire which was now continuing in an easterly direction at the Breakneck Road, the Carriage Road and finally at the new Eagle Lake Road, the latter being successful.

Before the danger, will continue until this section experiences a considerable rainfall.

Large crowds were busy both around the large area and, where possible, in the very fire itself attempting to wet it down sufficiently that it might be controlled.

It is impossible for one to name all the units and individuals who came forward unselfishly and unsolicited to take part in the battle. Canteen units from Northeast Harbor and Bar Harbor kept sandwiches and coffee on hand at all times, the latter being supplied by a large group of volunteers who worked day and night in relays. School teachers, nurses, businessmen, students, even printers pitched in to do their part. Apparatus from Brewer, Bangor and Old Town was sent here and used as stand-ins for the local department.

In the evening when it was reported the Lippincott Cottage at Northeast Harbor was ablaze, the Northeast Harbor apparatus necessarily withdrew and the Brewer apparatus with their crew was sent to aid and by a remarkable piece of work the fire was confined at that point which because of winds threatened many cottages of the Kimball House which was situated nearby. The Lippincott Cottage was a total loss, the fire having gained such headway before being discovered.

the traffic.

No words can express our prai[illegible] for the cooperation of the [illegible] departments, agencies, officials [illegible] volunteers during the crisis. At [illegible] present time no official estimat[illegible] have been made of the proper[illegible] loss or area devastated by the fir[illegible]

Bar Harbor Times
October 23

NOTICE

Supper Postponed

The baked bean supper at [illegible] Catholic Parish Hall has been po[illegible] poned one week due to the fir[illegible] The supper will be held Saturda[illegible] Nov. 1.

The Bar Harbor Times
October 23, 1947

Meeting Postponed

The Bar Harbor PTA meetin[illegible] which was scheduled for tonig[illegible] (Thursday), has been postponed.

The Bar Harbor Times
October 23, 1947

Weather Report

Week of October 17-23

Friday	62	Sunshin[illegible]
Saturday	65	Sunshin[illegible]
Sunday	68	Sunshin[illegible]

13.1 Bar Harbor Fire Anniversary supplement, the *Bar Harbor Times*, 19 October 1967.

to be on the alert for fire. Some people thought he was crazy. Others were too preoccupied with wells that had gone dry, fields which were too dusty to furrow, and cranberries and blueberries that needed to be picked. Hunting season was also underway.

By the second week of October, the woods were rated as Class Four, meaning that they were in a high state of flammability. The state's forest commissioner suggested that Governor Hildreth would be wise to close the forests until the rains came, just as the state of Vermont had done. Hildreth was too concerned that businesses would suffer, and that government revenue would be lost if anything serious was done to restrict bird hunting.

It was like the movie *Jaws*, playing out in the forest instead of the ocean. It was going to take more than few whiffs of smoke and dire warnings from the high-ranking forest commissioner and a lowly fire warden to get politicians to

13.2 The Bar Harbor Fire, 1947.

close the woods to hunters and campers, and to order farmers to stop burning their fields and residents to stop burning their trash. The state decided instead to keep the fire towers manned as a precautionary measure.

The consequences for procrastination presented themselves on 17 October when a woman called the fire department after seeing smoke rising from a cranberry bog not far from her home. The fire had likely been lit by a smoker who had dropped a cigarette or match while harvesting berries. The fire smouldered in the peat for several days before gusty winds fanned the flames. It was extraordinary to have so many men from the Army Air Corps, the Navy, the Coast Guard, the University of Maine forestry program, and Bangor Theological Seminary joining local firefighting crews. National Park Service employees were also flown in from parks throughout the east.

Formidable in number as these crews were, they had no chance of containing the fire once gale-force winds swept in and started shooting embers in all directions. Communication and coordination were serious problems, as always for any crew fighting a rapidly spreading fire. Chains of commands were at best confused with so many agencies involved. There was a shortage of hoses, pumps, and two-way radios, and not enough roads to give fire crews access to the hotspots. The only thing that worked well, mercifully, was the evacuation coordinated by the military, local police, fishermen, and the Red Cross.

What set this fire apart from others was that it was not just farmers, loggers, fishermen, and other ordinary folk who were being impacted. The rich and famous who had built million-dollar mansions and high-priced cottages at Bar Harbor had also been traumatized. Blaine Cottage, which was owned by Walter Damrosch, a famous composer and conductor of the New York Symphony, was obliterated, as was the home of novelist Mary Roberts Rinehart, the Agatha Christie of the United States. Only the chimneys of the elegant one-hundred-room Malvern Hotel, whose regular guests had included the fabulously rich and powerful, such as mining engineer Hennen Jennings and Gifford Pinchot's father James Pinchot, were left standing. Joseph Pulitzer's mansion *Chatwold*, which he bought in 1894, featured Bar Harbor's first heated swimming pool. It would have burned as well had it not been demolished beforehand in order to make way for a new building. The greenhouse and outer buildings, however, were destroyed. Pulitzer flew in on a private plane to survey the damage, which included "Woodlands," the summer home of his sister.

The fact so many of the mansions along Millionaire's Row burned did not go unnoticed by big city newspapers. Mary Van Rensselaer (Molly) Thayer, a society columnist for the *Washington Post*, was well ensconced in Boston–New York–Washington social circles. She devoted considerable ink to the story of the four Damrosch daughters, who wrote a cheque for $5,000 in aid of the homeless.

In the weeks after the fire, Governor Hildreth had a meeting with the wealthiest property owners to determine how the town was going to be rebuilt. (One of their goals was to have no paper companies operating in the state by 1947.) All of which is to say that when the rich and famous find themselves in harm's way, politicians who depend on their generosity and influence respond in rapid-fire fashion. These people had no need for the welfare cheques, food, clothing, and blankets typically doled out after a wildfire disaster. What they wanted was a system that would prevent this from ever happening again.

What followed was as momentous as it was extraordinary in the ways it would shape the future of forest fire management in North America. Gifford Pinchot, who had died the year before, would have been proud. An interstate compact was established later that year to promote effective protection and control of forest fires in the northeastern states and adjacent portions of Canada. The fact that Hildreth chaired the National Governors' Conference may have helped move the agreement along as quickly as it did. In any event, Congress approved the Northeast Compact in 1949.

The idea of Canada and the United States cooperating on wildfire detection and control had been tested back in 1937 when the province of Ontario and the state of Minnesota agreed to coordinate airplane patrols, radio and telephone facilities, and information from lookouts on both sides of the border.[3] But the Northeast Compact was much more formal, legally binding, and remarkably sophisticated in the manner in which personnel and equipment were to be deployed. Member states and provinces paid fees based on their share of the forests that initially included an area three times the size of Maine.

Original members included the states of Maine, New Hampshire, Vermont, Massachusetts, Connecticut, Rhode Island, and New York. Quebec and New Brunswick joined in 1969 and 1970, Nova Scotia in 1996, and Newfoundland and Labrador in 2007. It was such a good idea that the number of compacts in North America eventually expanded to seven. They now include every province in Canada and every state except for Nebraska, Utah, Kansas, New Mexico, Arizona, and California. The commission's mandate was and still is "to provide the means for its member states and provinces to cope with fires that might be beyond the capability of a single member through information, technology and resources."[4] The rules of the commission allowed for fact-finding and coordinating within a deliberative body that had the power to make recommendations on training and for all technical issues related to fire control.

Once the Northeast Compact was established under the guise of the Northeastern Forest Fire Protection Commission, little time was wasted in moving things forward. In 1951, member states sent firefighters and managers to the forest fire training school at the Yale School of Forestry Camp in Norfolk, Connecticut. Representatives from New Brunswick, Quebec, New Jersey, and Pennsylvania also attended. The participants got a first-hand look at a dangerous situation in New York state, where one hundred hectares of forest had been laid bare by a hurricane that had ploughed through the area the year before. It was a huge tinderbox waiting to be ignited.

Instructors included the usual list of speakers from Yale, and from state and national parks. But the presence of William Fuller, chief meteorologist with the United States weather bureau in Connecticut, was a sign that forest firefighters were, thanks to American forester Harry Gisborne and Canadians James Wright and Herbert Beall, increasingly interested in how drought, wind, humidity, and hot weather influenced fire behaviour. The participants even got to successfully fight a 2,500-hectare fire on Great Mountain in northwest Connecticut. This one was imaginary, albeit "grimly realistic in every detail except for the smoke and flames," according to the *New York Times*. The goal was to prevent the so-called "Fire Dog" wildfire from doubling in size and crossing state lines. Racing against time, the forty firefighting experts from both countries put their heads together to mobilize interstate personnel and resources.[5] By all accounts, they succeeded.

The Northeast Compact got to engage in a real-life scenario the following year when seventy fires in Connecticut overwhelmed the best efforts of more than 1,000 forest rangers, city firemen, and volunteers. It was the first of many future collaborations. In the US, the establishment of the fire compact made up for the loss of the Civilian Conservation Corps. In Canada, it gave intellectual and physical muscle to a system in which most provincial firefighters were poorly paid, mostly seasonal, and constantly in fear of losing their jobs. No longer would it be acceptable for the training of these men to be as inadequate as it often was outside the national parks. Canadian provinces could no longer, in good conscience, employ "misfits," as General Kennedy's royal commission described some who fought fires in Ontario. And they could no longer rely simply on rangers on horseback or in canoes, talented as they may have been at carpentry, tree-felling, compass reading, bookkeeping, cooking, and cleaning. There was just too much influence and integrity, not to mention money, associated with being a member of a compact to allow a province, state, or the federal government to short-change the system.

Had Ernest Finlayson been alive, he would been heartened by the establishment of the compacts and the influence they had on decision-making in Canada. In 1948, the National Research Council began working with the Ontario Lands and Forests Department to experiment with ways of making it rain over a fire. The passage of the Canadian Forestry Act the following year may not have had anything directly to do with the establishment of the Northeast Fire Compact that same year. But it, and a funding deal made with Alberta two years earlier, brought more money to the table, which was essential if the Canadian provinces were going to stand shoulder to shoulder with the Americans on fire lines, in fire training programs, and in the sharing of technology and equipment.

Most of the money went to inventories, reforestation, fire protection, the construction of access roads, and for a frightful plan to develop a virus that would kill the spruce budworm, which was infecting millions of trees in eastern and central Canada. According to a *Globe and Mail* report in September 1950, the insect laboratory that was testing it out was up and running and requiring everyone who went into the culture lab to shower and wear protective clothing.[6] But there was also funding that would eventually allow the Canadian Forest Service, as it had come to be called, to incrementally hire new staff, build new labs, and set up regional offices across the country. The expanded research role that Finlayson was aiming for when his organization was forced to abdicate forest protection and management duties was now within reach.

Scientists and engineers from both countries had developed many new gadgets and contraptions for predicting and detecting fires. Harry Gisborne, who died in 1949 on his way to inspect the site of the deadly Mann Gulch Fire in Montana, invented the Asman aspiration psychrometer (measures humidity), visibility meter (measures air clarity), anemohygrograph (measures wind speed), double tripod heliograph (used to flash signals with sunlight), and the blinkometer (used to assess moisture conditions). J.E. Reid of the University of Toronto had some success with a "bolometer" he invented. This was an optical system that scanned the landscape looking for hotspots and transmitting their location without it being necessary for anyone in a fire tower to be on hand.

The one thing that distinguished fire research in Canada from fire research south of the border was the means by which goals were reached. Home-grown scientists such as Charlie Van Wagner, a chemical engineer who joined the Canadian Forest Service in 1960, thought that the "answers to predicting fuel moisture and fire behavior would never be found in expensive laboratories or through sophisticated complex maths and physics,"[7] as the Americans insisted. He and other Canadian fire scientists believed that the best approach was the kind of field work that started in the 1930s with two-minute test fires lit by a match, and then much larger experimental fires that began in the 1960s. Van Wagner was a pioneering member and one of the technical leaders of the federal forest service's "Fire Danger Group." This group included John Muraro, Jack Turner, and Bruce Lawson from the Pacific Research Centre; Dave Kiil and Dennis Quintilio from the Northern Forest Research Centre in Edmonton; Al Simard and Dave Williams from the Forest Fire Research Institute in Ottawa; and Brian Stocks and John Walker from the Great Lakes Forest Research Centre in Sault Ste Marie.

All these people had analytical skills and extensive experience working in the field. Muraro, a former smokejumper who many say was the best field scientist among them, was the first to experiment with large experimental burns, such as the 160-hectare prescribed fire he helped orchestrate in a logging slash in the interior of British Columbia in 1963. He and his colleagues also produced airborne ignition machines such as the helitorch, and automated fire weather stations to collect and transmit data from remote sites to central offices. Once commercialized, these station networks advanced local weather forecasting to a new level, improving the prevention and suppression of wildfires and prescribed burns.

The group's effort to turn the Wright-Beall index into one that would take into account the varying range of fire weather in Canada and the sixteen different types of fuel that are most commonly found in the country's northern forests succeeded well beyond the expectations of administrators when it was instituted across Canada in 1972 as the Canadian Forest Fire Weather Index System. Increasingly sophisticated iterations of the index evolved with the best of the old features preserved and new ones incorporated.

There was, of course, competition between Canadian and American fire scientists. But much of it tended to be good-natured, because many Canadian fire scientists, like Muraro and Kiil, got their graduate degrees at the University of Montana. And several American fire scientists, including Marty Alexander, joined the Canadian Forest Service during its days of expansion. The challenge of predicting, detecting, and suppressing wildfire in both countries grew into an international effort, successfully in most cases but with some unintended and unexpected consequences. In fact, Canada and the United States got to be *too* good at suppressing what Mother Nature intended to do.

14

BLUE MOON AND BLUE SUN

On 25 September 1950, "a weird red smog" estimated to be 600 kilometres long, 500 kilometres wide, and five kilometres thick blanketed most of southern Ontario, parts of the American Midwest, and most of the northeast region of the United States. Canadian astronomer Helen Sawyer Hogg confessed that she had never seen anything quite like it. "The western sky became a dark, terrifying mass of cloud and haze, as though a gigantic storm was approaching," she wrote in the *Journal of the Royal Astronomical Society of Canada*.[1] Police switchboards were overloaded with calls from frightened people who had heard rumours that an atomic bomb had exploded. Some believed that a third world war had started. As in the "Dark Day" of 1780, many went to church to pray, believing that the return of Christ was at hand.[2]

Northwest Airlines reported that one of its pilots had radioed in warning that he would be forced to land because of poor visibility. Flying over Michigan, the pilot reported smoke seeping into the cabin, frightening passengers who thought the plane was on fire. Instead of resorting to an emergency landing, American Airlines pilots soared to great heights in a futile effort to get beyond the pall. One, flying between Cleveland and Pittsburgh, had never experienced anything like this before. He too thought that the plane was on fire.

It was not an alien invasion, a volcanic eruption, or an eclipse of the sun, as others suspected. But in places such as Toronto, Pittsburgh, Cleveland, Fort Erie, and many towns in New York state, it was so dark at midday that the lights at baseball fields, including those at Yankee stadium, had to be turned on to illuminate play. No one in Buffalo knew what to make of the gloom when it started snowing shortly after the streetlights were turned on mid-afternoon. And in an article in a Jamestown, New York, paper, a farmer described how his chickens, which had fanned out for their midday foraging, "suddenly realized they were being caught by darkness, so they scurried back across the cow yard in more than usual earnest, their heads moving in delayed jerks."[3]

The pall that the fire's smoke cast over eastern North America generated so much public concern that the *New York Times* put the story on the front page.[4] In that article, the effects of the fire were compared to the eruption of Mount Krakatau, which killed 36,000 people and sent a fine ash around the world after it erupted in 1883. Many New Yorkers who didn't have access to big city papers didn't know what to make of the event. A Mrs Dora Gesaman of Watts Flat "announced gravely that at four p.m. when the overcast lifted, how her rooster crowed as if it were dawn and the chickens left the roost under the impression that a new day had arrived."[5] One elderly man in the town of Busti in New York State was so unnerved that when a relative went to check on him, he was shaking like a leaf. "Do you think this is the end of the world?" he asked.[6]

Some radio stations didn't help calm fears. One announcer told its listeners that "Canada was on fire." When Canadians heard the rumours, they called radio stations in southern Ontario asking whether they were true. "Everyone remembers what he was doing when he heard that President Kennedy had been shot, that Pearl Harbor was bombed or that either world war had ended," local historian Norman Carlson wrote in the Jamestown *Post Journal*. "So too everyone my age and older remembers another event: a Sunday afternoon in 1950 when the sun ceased to give her light and our primitive fears of darkness, mortality and powerlessness rose at least near enough to the surface to etch a lasting trace that belied our calm."[7]

The mysterious origins of the pall of smoke led to a number of conspiracies theories, as Canadian fire scientist Cordy Tymstra describes so well in in his book on the topic.[8] Some, he noted, thought it was the work of the Central Intelligence Agency. Others speculated that it was a seed-clouding experiment that the government badly botched and was covering up because it was designed to provide cover for bombers. Many thought it was produced by an accidental nuclear explosion.

The beginning of what many people thought was the end of the world took place on 2 June 1950, when a small fire ignited in northern British Columbia near the Alberta border. It had been an exceptionally hot spring, and forest firefighters were too busy battling other fires in British Columbia, Alberta, and the southern Yukon to do anything about a little one far from human settlement. Outside of national parks, the policy back then was to ignore fires that were ten miles distant from roads, power lines, villages, or mine sites. An

14.1 In the early days of fire suppression, rangers rode horses into the backcountry.

extremely remote fire, such as this one more than sixty kilometres north of the Canadian National Railway, wasn't in the playbook. It was a "ghost fire." As far as forestry officials were concerned it did not exist, except to local forest ranger Frank LaFoy who, like many of his colleagues, used boats and packhorses to patrol the countryside because there were few roads to drive on.

Within a few days, the "ghost fire" danced across the Alberta border into the largely uninhabited wild lands of the Chinchaga region. Fuelled by a tinder-dry forest that went on forever, the relatively small blaze developed into a wildfire of monstrous proportions. When a Royal Canadian Air Force pilot flew over the region on 21 September 1950, he counted more than a hundred fires burning what ended up being two million hectares before wet and cold weather put them out.

The Alberta Forest Service's small fleet of bulldozers, and teams of a dozen or so firefighters who had nothing but "piss packs" – hoses connected to water packs on their backs – couldn't fight fires as big as these. It was like using a squirt gun to extinguish several flaming city blocks of buildings.

The pall of smoke crossed the Atlantic where it created a "strange blue sun followed by a freak blue moon in the skies over the North Sea." British stargazers were "baffled by the phenomenon," according to the *London Times*. "Jittery old folks queued up at some provincial Danish banks demanding their money for a hurried flight from doomsday … A suspicious few feared the end of the

14.2 Firefighter using a hose pack to extinguish fire. Photo by Felix H. Man.

world," the paper noted. The *London Times* also took note of the blue moon as well as the Canadian fires, putting both on the front page next to each other.[9] The editors, however, failed to make a connection between the two events. The *Boston Globe* and other papers did so with remarkable insight. A reporter for the *Globe and Mail* harkened back to 19 May 1780, when a similar phenomenon had darkened the skies of New England. People were just as concerned then, the writer noted, most believing that the end of the world was at hand. Meteorologists, however, offered other explanations. One likened the unearthly pall to a lump of mud being dropped into a clear stream. "A streak of mud will follow the current."[10]

Southern Ontario was among the regions hardest hit. The sudden dusk at midday added 180,000 kilowatt hours to local power consumption, almost equal to that being produced daily by a new power plant on the Ottawa River. Locally, the heat was so intense that soil chemistry was altered so that tree regeneration could no longer be supported. Communication between Alaska, the Yukon, and the rest of the world was disrupted for several days when the lines went down. "All over the world, meteorologists were puzzled," stated an article in the *Globe and Mail.*

The Chinchaga Fire of 1950 burned for 222 days and torched a stretch of forest that was 175 miles long. It was and remains one of the biggest conflagrations in North America since the Miramichi Fire of 1825. More than 10,000 square miles (2,589,988 hectares) of forest went up in flames. No one died as a result of this fire. What made it extraordinary was not only its size, massive as it was, but the breadth of its impact. The Chinchaga Fire created the world's largest blanket of smoke. Not only was that smoke transported halfway around the northern hemisphere, it caused both the moon and the sun to appear blue. No other fire of any magnitude has been known to have done that.[11]

Chinchaga, however, was just one of hundreds of fires that burned forests from the sub-Arctic of North America to New Mexico that year. Western Canada was so hot and dry for so long that farmers in Saskatchewan worked around the clock in late September digging trenches around wheat fields that had not yet been harvested. "The devil," one farmer told a reporter from the Canadian Press, "has moved in with us."[12]

In New Mexico, the Capitan Gap Fire torched just 7,000 hectares of the Lincoln Forest. The damage, however, was substantial, and while soldiers searched through the charred remains for any sign of life, they found a badly singed bear clinging to a tree. Nursed back to health, "Smokey Bear" became the real-life symbol of a fictitious animal that the US Forest Service had created a few years earlier in order to educate Americans about the dangers of forest fires.

What the Chinchaga did was what the "Dark Day" fire of 1780 portended. Conflagrations like this would no longer be ghost fires, burning innocently through the boreal forests. As oil and gas exploration and timber harvesting delved deeper into the boreal forest, there would soon be too many people and too many assets standing in harm's way to conduct light burning exercises or to allow lightning-triggered fires to burn themselves out. With the intersection between the forest and the urban industrial landscape on a collision course, scientists in Canada and the United States started to think boldly about what might be done to cushion the impact.

15

NUCLEAR WINTER

In 1983, astronomer Carl Sagan, atmospheric scientist Richard Turco, and three others wrote a paper in the journal *Science* describing what might happen if the Soviet Union and the United States engaged in an all-out nuclear war. According to their "nuclear winter" scenario, the explosion would result in massive forest fires raging across the continent. Smoke from these fires and from fires burning in cities would be injected not just into the lower atmosphere, where it would eventually be removed by rainfall, but all the way into the stratosphere, where it would linger for years. There would be "quick freezes": very rapid temperature drops due to plumes of smoke rising above the troposphere. Sunlight would be obscured for weeks, months, or possibly even years. Temperatures over the longer term would cool to levels too low for farmers to successfully grow crops. Millions of people would starve. Many more millions would be stricken with radiation sickness.[1]

They were not the first to come up with this theory. Paul Crutzen and John Birks had made a less dramatic case earlier when they wrote their now famous "Twilight at Noon" article in the journal *Ambio*.[2] A companion study done by Stanford University biologist Paul Ehrlich, Harvard University biologist Stephen J. Gould, Sagan, and many others went farther, suggesting a nuclear winter could lead to the extinction of many of the Earth's animal species, including the human race.[3]

Sagan is rumoured to have contacted the Canadian Forest Service about the Chinchaga Fire while he was conducting research for the scientific paper that he, Turco, and the others were writing. Brian Stocks, head of the Canadian Forest Service's Fire Research Program, was aware of the nuclear winter theory when he attended a meeting with US Forest Service colleagues in Missoula in 1985. Turco was there along with several other Defence Nuclear Agency–sponsored scientists. When they learned that Stocks was conducting large-scale prescribed burns not unlike those that might burn during a nuclear winter, an animated conversation began that ended with Stocks inviting them to come to northern Ontario to see for themselves what an intense fire looked like. He and colleagues from the Ontario Ministry of Natural Resources had planned to light fifty fires that summer.

The timing was not the best. Many Canadian nationalists were angry that their government was allowing the US to test cruise missiles in the north. Turco's visit also coincided with the US *Polar Sea* icebreaker sailing through the Northwest Passage, thumbing its nose at Canada's claim to Arctic sovereignty by purposely failing to ask permission that would have, in all likelihood, been granted without fuss. Meanwhile, Canada's Royal Society was in the process of distributing its own very grim appraisal of the impacts that a nuclear winter would have on the country's environment

Stocks and his colleagues were getting more comfortable setting bigger and bigger areas of forests on fire both for management purposes and to better understand fire behaviour. The site they took the nuclear scientists to was near Chapleau in northern Ontario, where thousands of trees had been killed by the spruce budworm infestation. The Chapleau Experimental Burn was first set in the centre of the area by a helicopter with a torch slung underneath. It then lit a path of flames in concentric swings toward the area's perimeter. The updraft that the centre fire created drew in flames from the outlying areas, throwing up a massive smoke column from which the scientists could get an idea of how the airborne detritus from a nuclear conflagration would behave. The cloud covered an area ten kilometres wide and forty kilometres long.

Turco was impressed. "Potentially, there is a lot that can be learned from this," he told newspaper reporters shortly after the experiment was completed. "I can see some potential for doing some future work. I'm more optimistic now than I was before I came here." Government officials knew about Turco being there, but they weren't happy about the press coverage that followed. Instead of reporting on the importance of lighting fires like this to better understand fire behaviour, the *Toronto Star* put the story on the front page on 4 August with an alarming headline that screamed "Controlled Fire to Test Nuclear Theory."[4] Other papers could not resist doing the same.

Stocks was mortified. "From that point on, this routine burn quickly became a nuclear winter fire," he recalled. "The US scientists were asked many questions, none about prescribed burning. I ended up on CBC's *As It Happens* radio show the

15.1 The Red Lake Experimental Fire, 29 May 1986. Photo by Brian Stocks.

next day trying to tell the whole story, but no one was interested by then. It was all about Canada playing a role in nuclear winter."

Fuelled by this misinformation and the subsequent public outcry over allowing what many wrongly saw as Americans conducting nuclear winter tests in Canada, the Ontario government closed the door to more collaboration. But then things started to settle down, as paperwork and background briefings made their way to ministerial offices. Canada needed experimental burns such as these to better deal with climate change, spruce budworm die-offs, and smoke chemistry issues. Nuclear winter was only a small part of what was going on near Chapleau.

The following year, the Ontario government had a change of heart. A minister in the government quietly asked Stocks to re-engage with the nuclear winter scientists. Stocks was happy to oblige because he and his team were interested in leveraging funding from the Defence Nuclear Agency so they could study mass-fire ignition dynamics from an operational safety standpoint. It was the beginning of another kind of Fire Compact between Canada and the United States, albeit one that came without dues having to paid by Canada. Very quickly, Stocks and his team got access to specially equipped aircraft and other high-tech tools that they could never have dreamed of attaining through federal or provincial programs.

"The Defence Nuclear Authority funded a number of subcontractors in the US, along with the US Forest Service, NASA, NOAA to come to northeastern

Ontario," Stocks told me in one of several interviews I did with him. "We conducted aircraft and ground measurements of fire behaviour and smoke chemistry in a way we never could with the budget we had within the Canadian Forest Service. Between 1987 and 1991 we documented close to 10 burns." This was the beginning of international, interdisciplinary fire research. Budget issues forced Canada in that direction, but the research was also made necessary by global awareness of fire as it related to atmospheric chemistry and climate change. Along with the United States, Canada had become a world leader in understanding fire behaviour, and the rest of the world was increasingly in need of that expertise. "Nuclear winter was a short-lived bit of serendipity" recalls Stocks. "I was not happy about the publicity at the time, but it worked in our favour over the long run, in spite of the fact that the Canadian government was unwilling to invest in fire science.

The "nuclear winter" theory lost traction when scientists concluded that the most likely targets were too far from forests that were big enough to fuel anything close to a nuclear winter. Even if that happened, others argued, the trunks of the trees, which contain 80 per cent of their biomass, would have to be reduced to ashes. In addition, the inner part of a tree's trunk absorbs more heat than the outer layers produce. This phenomenon would diminish the impact of thermal radiation.[5]

The spectre of an intense, all-encompassing fire much bigger than the Chinchaga Fire, however, presented itself throughout the 1980s. During 1982–83, bushfires in Australia consumed 520,000 hectares in what was up until then the country's worst fire season since 1939. In 1983, large fires burned 3.6 million hectares of tropical rainforest in the East Kalimantan region of Indonesia. Four years later, the Black Dragon Fire burned 1.33 million hectares in China. Many of these blazes behaved outside the bounds of what had been predicted by modern methods and spread much faster than fire behaviour models anticipated. But the Black Dragon Fire that raged through the remote Heilongjiang province of northeast China stood out from them all.

Nothing like it had been recorded in the annals of twenty-four dynasties.[6] In fact, there was no human record of a fire burning so much timber anywhere in the world. Entire towns were wiped out as walls of flames thirty-three metres high, driven by 100-kilometre-an-hour winds, ripped through everything in their path. At least 220 people died and 33,705 people were left homeless. At the height of the blaze, 60,000 soldiers were trying to slow it down.

As it turned out, the Canadian angle to this story involved Stocks, and China's need for Canada's expertise. The story got its start in 1973 with Canadian prime minister Pierre Trudeau's landmark visit to Beijing to commemorate the renewal of diplomatic relations between the two countries. The meetings that took place not only opened diplomatic channels between the two countries, they set the stage for several exchange programs, including one intended to capitalize on Canada's growing expertise in fire prevention and management. Concerned about fires that had been ravaging the forests in Heilongjiang, China signed an agreement with the Canadian International Development Agency and the Ontario provincial government to provide technical expertise on fire management over a five-year period.

Stocks got involved in 1985, shortly after Turco and the nuclear winter scientists went home. He spent six weeks in northern China that summer with the goal of establishing a modern fire-management project in Heilongjiang Province. "I was there to evaluate the suitability of the Canadian Forest Fire Danger Rating System for use in this heavily-forested region of China," he told me. "We worked very close to the Russian border (Amur River). This area was part of the Great Dragon Fire eighteen months later."

The Black Dragon Fire began on 6 May when, allegedly, an eighteen-year-old man spilled gasoline while fuelling a brush-cutting machine. There was little that could be done to put it out. The region was in the midst of a severe drought, and Canadian and American know-how was just beginning to be assimilated into Chinese fire management regimes. Many of the soldiers who were brought in had no tools whatsoever to fight the fires. Often, they soaked their tunics in water, beating the flames with them as best as they could

Canada offered to send water bombers the third week of May, before both governments decided that they would not be able to get to the region in time. However, China accepted 2,000 backpack pumps, 400 fire shovels, one hundred fire axes, and one hundred pump repair kits, all of which had to be transferred to Chinese planes because the Canadian aircraft were deemed to be too big to land on runways in northern China.

The fire lasted for twenty-seven days. But it was one day of "phenomenal fire behaviour" that did the most damage according to Stocks, who conducted a post-mortem. "On the second day, the fires spread at an average rate of six to eight kilometres per hour throughout the afternoon and evening, covering more than 350,000 hectares. We documented this with satellite photography and in collaboration with Chinese fire control personnel who were on site. The die was cast on that one day. Suppression would be long and very difficult."

What really struck Stocks was that there was nothing in the North American experience to explain how the fires spread so quickly. If this was the future of fire, he wondered, how bad might things get?

16

YELLOWSTONE: A TURNING POINT

A couple of hundred metres shy of the 3,000-metre summit of Fawn Pass in Yellowstone National Park, a swirling mass of snow drowned out the last rays of sunlight, forcing my companions and me to stop our trek across sweeping drifts deeper than our knee-high gaiters. There was no chance of getting beyond the pass, where we needed to be that night if there was to be any hope of arriving at Fan Creek trail in two days, as we had planned.

As we set up camp on a lower plateau beneath the canopy of spruce trees, the weather worsened. Moist gobs of snow piled on my single-person tent, then slipped down the sides, jolting me upright at one point in the mistaken belief that a grizzly bear was trying to get to me. When morning arrived, the few patches of green that were visible on the ground the night before had been transformed into a scene more like Christmas Eve than the second week of June. Realizing that I did not have the time in my tight schedule to spend another night or two on the mountain waiting for the weather to clear, I said good-bye to my partners, wishing them well.

On the way back down, I had another one of those heart-stopping moments when I passed steaming piles of bear dung. Just ahead, fresh tracks in the snow suggested that a big male was following a female and a cub. As I stopped to scan the hillsides around me with my binoculars, I spotted a dozen elk standing stiffly on a bald, wind-blown hilltop that had burned to a crisp when the Fan Fire ripped through the region in late June and July of 1988. Heads erect, ears twitching, the elks' eyes were trained on something below. Two ravens entered the picture, flushing out some songbirds that had taken refuge in the bushes. My inclination was to stay where I was until the more seasoned part of my psyche reminded me that I had a whistle and some bear spray, and that I was on a trail with a good view, thanks to that fire in 1988. Grizzly bears, I knew, would rather flee than confront a person who was not threatening them.

Minutes after gingerly crossing a narrow log bridge over the wild, crystal-clear waters of the Gardner River, there was a sharp turn on the trail that gave me a close look at what was spooking the wildlife. No more than one hundred metres away in a meadow, a female elk was racing back and forth like a horse in a coral trying to avoid a lasso. Only it wasn't a rope that was freaking her out. It was two grey wolves, snapping at her legs and jumping at her throat.

Transfixed, there was nothing for me to do but watch in silence as this drama unfolded. It didn't take long. Sensing that she was doomed if she stayed in this situation, the bloodied elk made a run for it. The last I saw of her, she disappeared into a cluster of aspen and knee-high pines that had risen from that fire where old growth conifers once dominated. The wolves were in hot pursuit.

The Fan Fire was the first of many mostly lightning-triggered fires that burned a third of Yellowstone. Like the Big Burn of 1910, the 1988 fire season was instrumental in reshaping public policy and scientific understanding of the role that fire plays in the ecosystem. Until around 1965, national parks officials were just as diligent in supressing fires as they had been in removing wolves from the landscape decades earlier. The last wolves in Yellowstone were killed off in 1926. With some exceptions, the 10 a.m. rule in Yellowstone was as religiously adhered to as it was in Yosemite and other national parks.

It took some time to make the connection, but park naturalists and scientists working in the park eventually realized that the removal of wolves had led to a plethora of elk and moose that feasted heavily on the park's aspen and riparian willows. Without fire, pine and spruce came to dominate the ecosystem in ways that did not favour the root vegetables and berries that black bear and grizzly bears depend upon. Fewer animals were being killed because there were no longer many large predators. The less carrion there was left behind by these animals, the less eagles had to eat. Scientists were seeing similar, worrisome trends in other parks such Yosemite, where the Giant Sequoia stands were getting older, not younger. There were a lot of magnificent trees that were a thousand, even 2,000 years old. A few were even older than that. Seedlings, however, were not taking root. And in the Everglades, slash pine and piney vegetation was giving way to broad-leafed hammock and other hardwoods. Banff National Park was thickening up with spruce and pine that were seventy to one hundred years old.

In the 1950s, scientists such Stanley Cain, a botanist from the University of Michigan, suspected that the absence of wildfire on the landscape was the reason these ecosystems were changing so drastically. But it was difficult to convince the cadre of park managers who had spent a lifetime being told, and in turn telling millions of tourists and backcountry adventurers, that fire was a demon. They were more interested in preserving the atmospheric splendour of the parks for a rapidly rising number of visitors, and the revenue they generated, than maintaining ecological integrity. Wolves were as bad for elk as fire was for forests. Uniform stands of eighty- to one-hundred-year-old spruce and pine trees were preferable to a fire-scarred landscape in which aspen, willow, and fireweed proliferated among charred stumps of burned wood.

Not everyone in the national parks and US Forest Service system was brainwashed into this way of thinking. Even in Pinchot's time, there were people like Yale graduate Herman Chapman, who joined the US Forest Service in 1904. He worked for a time in the south where "woods burning" was common and effective in regenerating forests. As early as 1912, he held the notion that fire could be used as a tool to better manage forests everywhere. "The attempt to keep fire entirely out of southern pine lands might finally result in complete destruction of the forests," he wrote in an article in *American Forests* that year.[1]

Elers Koch, a well-respected forester who was on the front lines of the Big Burn of 1910, thought so as well, albeit for other reasons. He saw the building of roads into the wilderness to better fight fires as yet another assault on nature. He firmly believed that the money being spent to fight fires had made no appreciable difference in the amount of area burned.[2] When he wrote a landmark essay in the *Journal of Forestry* questioning the economic cost of fighting fire and making the case for letting forested wildlands burn for ecological reasons, it was summarily dismissed by an accompanying article written by Earl Loveridge, the man who came up with the 10 a.m. policy.[3] Loveridge was then the assistant chief of the Division of Operation and Fire Control at the Forest Service's Washington office.

No one was willing to champion the need for light burning so long as the likes of Pinchot, Graves, Greeley, Stuart, and Silcox served as forest chiefs. Their goal was to slay those fire-breathing dragons in the forest. It was a view shared by those who ran the national parks services in both the United States and Canada. Yet, in time, some US park officials suspected there was an element of truth to what Chapman and Koch preached. They came to recognize this as plans for a national park in the Everglades were being made in 1930s and 1940s. Fire had always been part of the soggy landscape in southern Florida, when drought extended its reach into the woods, prairie, and tree islands that moved about on floating peat. But it came to be an increasingly powerful force following decades-long efforts to drain the Everglades for cities, farms, and airport runways. In 1937, fire burned half of the piney woods, 80 per cent of the Everglades sawgrass prairie, 30 per cent of the coastal prairie, and 5 per cent of the Ten Thousand Island coast.[4] In 1945, "there was no water to be found anywhere in the glades," according to a park report.[5] Fires burned well into June that year.

A crisis of sorts came into play in 1950, when three big fires in the Everglades National Park burned simultaneously, overwhelming the best efforts of firefighters to put them out. As more plans were put forward to drain and divert even more water from the region, Daniel Beard, the superintendent of the Everglades from 1947 to 1958, realized that fire was the only way to prevent hardwoods such as broad-leafed hammock from creating a dense understory that would result in fires so intense they would result in the extinction of slash pine and other pineland plants. His proposal to burn in a controlled fashion, rather than one dictated by nature or careless humans, was inspired by William Robertson Jr, whose research in the Everglades in the early 1950s highlighted the need for prescribed burning. South Florida's vegetation is what it is because of fire, according to Robertson, and the only way of allowing it to persist was to let it burn naturally or bring back fire with prescribed burns. Bear's plan to burn made its way up the chain of command. It was accepted in 1957 by NPS director Conrad Wirth. This led to the first controlled burning in a national park in thirty years.[6]

These well-founded concerns about ecological integrity in national parks resulted in a government inquiry chaired by zoologist A. Starker Leopold, son of the legendary conservationist Aldo Leopold. The original purpose of the inquiry, which included Stanley Cain, was to assess the highly controversial practice of shooting elk in parks like Yellowstone where there were no wolves to control their numbers. (Park officials in Elk Island National Park in Canada were doing this as well.) The inquiry was later broadened to include fire and other ecological issues. The landmark recommendations made in 1963 set the stage for a new approach to wildfire in national parks. Biologists, rather than professional foresters, were given priority in a new wave of hiring. Their insights led to an interdisciplinary approach that allowed for some lightning-triggered fires to burn themselves out. It also led to a series of experimental burns that were to be evaluated for management purposes.

The success of those burns convinced the US National Parks Service in 1972 and the US Forest Service in 1978 to use prescribed burning as a management tool. The development of this new relationship with wildfire in the forests, however, was fraught with trial and error, as well as pushback from politicians, the public, and from some plant and animal ecologists who continued to see fire as a destroyer and not a birther. The biggest setback occurred in 1988, when Yellowstone burned in a way that horrified decision-makers and rocked the wildfire world in a way that had not happened since the Big Burn of 1910.

The fires began naturally when lightning triggered fires in the same remote area of the park I hiked through twenty years later. Managers initially allowed the fires to burn, as they had done a number of times since 1972 when the park embraced the "let it burn" strategy. Since that time, lightning had been responsible for 300 fires, 80 per cent of which burned less than an acre before dying out. In the decade before this fire, only 12,000 hectares of Yellowstone's forest had been consumed by fire. Letting these remote fires go seemed like a safe bet.[7]

Computer models that predicted fire potential in the park backed up the plan. The summer of 1988, however, turned out to be the hottest and driest in 110 years. Lightning that came with the extreme heat struck tinder-dry timber. Sparks from chainsaws used to create fire breaks ignited more flames. The conflagrations quickly spun out of control, overwhelming all efforts to quell them once the decision was made in mid-July to close most of the park and send firefighters in.

It was as weird a scene as the park had been witness to in its 116-year history. Members of a Christian camp based just outside the park refused to leave, instead holding hands and chanting in the hope that God would intervene and bring rain. Tourists continued to arrive despite the closures and the evacuation of Yellowstone headquarters at Mammoth Hot Springs. Many went home with t-shirts that called for more firefighters because the bears liked eating them. Some people drove in just to see the spectacle. In spite of all that had happened in the past, wildfire continued to exist outside the knowable because there was still not enough science to explain it, nor any long-term memory of it. The public continued to believe that if there enough firefighters on the ground and water bombers in the air, fire could be controlled.

The ghost of Gifford Pinchot must have been haunting reporters who had bought into Smokey Bear ads, wartime forest fire prevention posters, the scourge of Japanese fire balloons, Richard Turco's and Carl Sagan's nuclear winter theory, and the policy which specified that all fires be brought under control by 10 a.m. the morning after. How, they asked, could this been allowed to have happened?

Like the Statue of Liberty, Yellowstone was an icon that represented the best of what America symbolized. For most reporters covering the event, it was hard not to think that some big mistake was made as troops were brought in to help 3,000 firefighters save buildings around Old Faithful, but ultimately lose the fight to save magnificent old growth forests.

They were not alone. People were furious when they learned through daily television and newspaper reports that officials had allowed lightning-triggered fires to burn in the park. A "public trust" had gone up in flames. "The idea that to let a forest burn down to the ground is somehow good for it, is akin to the concept that saturation bombing is good for urban renewal because it levels all structures," wrote one reader in a letter that was typical of many published by the *Los Angeles Times* and other papers. The reader was not alone in calling for the immediate resignation of all park officials who were responsible for the "let it burn" strategy.[8] One Wyoming rancher echoed the view of many when he suggested that Yellowstone was going to be a "black forest for 100 years." The

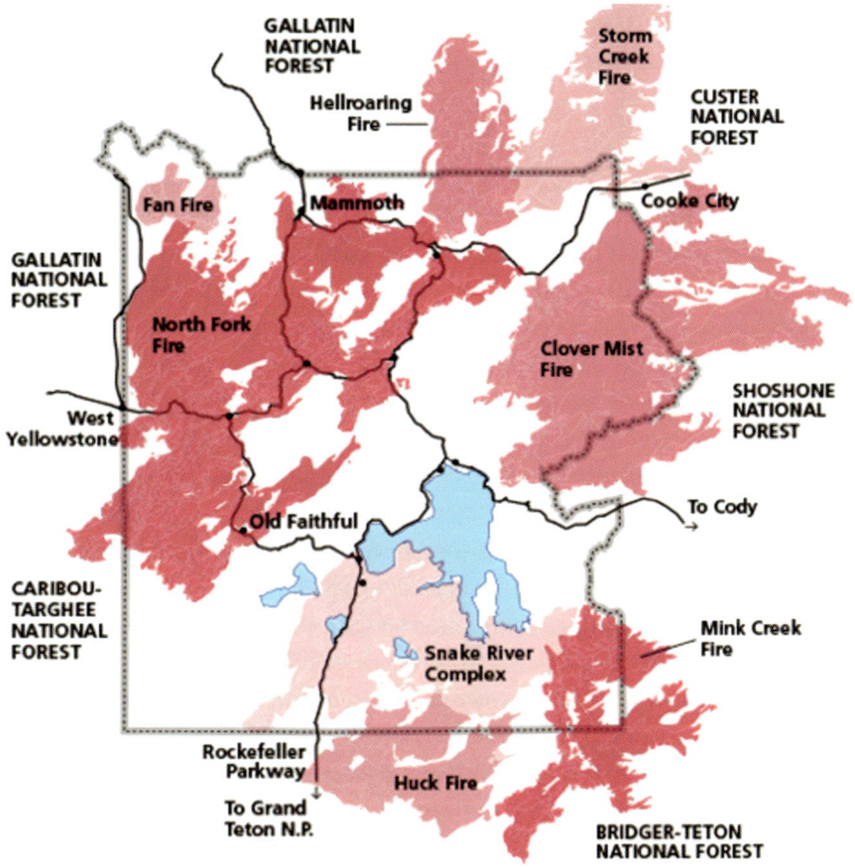

16.1 Map of fires in Yellowstone from 1988.

head of the American Forest Products Association also argued that to let it burn had been a bad idea.

It didn't help park officials when President Ronald Reagan declared Yellowstone a national disaster area, while dismissing the "let it burn" strategy of the national parks and forest service as "a cockamamie idea." At his behest, Interior Secretary Donald P. Hodel, Agriculture Secretary Richard E. Lyng, and Deputy Defence Secretary William Howard Taft travelled to Yellowstone to view operations and to meet with the public. Hodel vowed that he would never allow something like this to happen again, but stopped short of calling for the resignation of William Penn Mott Jr, the director of the national parks service.[9] That didn't stop Wyoming's two Republican senators from doing just that. Senator Malcolm Wallop was especially angry, describing the "let it burn" strategy as "absurd" and "scientifically indefensible."[10] The attack on the policy was, unfortunately for parks officials, a bipartisan one, which was rare given how often the Reagan administration's policies on environmental issues were ridiculed. "They will never go back to this policy," declared John Melchor,

Democratic senator of Montana, who announced that he would hold hearings. "From now on, the policy will be putting the fire out when they see the flames."

Once again, politics tried to square the fire triangle. Mott had worked for national and state parks since 1933 and was well respected by parks service staff, who felt threatened by Reagan's assault on conservation imperatives. Mott held his ground in supporting the "let it burn" strategy even when an inspector general was brought in to investigate him. He didn't flinch when Hodel told *Nightline*, ABC's flagship nightly news program, that "it is clear that this 'let it burn' strategy has been a disaster in this year of fire."

Yellowstone superintendent Robert "Bob" Barbee refused to resign, even as the attacks on him took on what he called a "lynch mob mentality." Barbee insisted that he and others in the national parks and US Forest Service, which was fighting fires outside Yellowstone, had done nothing wrong; that it was weather, not the "let it burn" strategy that accounted for such an intense number of fires. "There is this incredible myth that somehow the Parks Service and the Forest Service are standing by, gleeful over what is a major catastrophe, because we have a policy that recognizes the role of fire in natural systems."[11] Barbee later told the *New York Times* that he was hoping that Libyan president Muhammar Gaddafi would do something outrageous to get him out of the media spotlight.

Following the fires of 1988, the secretaries of the departments of Agriculture and of the Interior established a multiagency Fire Management Policy Review, which concluded that the policy of allowing fire to play a role in wilderness resource conservation was sound. What was needed was a better way to mitigate the risks.

The national parks and the forest service did go back to burning, but not with gusto. In 1988, the year that Yellowstone burned, fifty fire plans in the 387 designated wilderness areas on National Forest System lands were approved. In 1992, only thirteen fire plans were operational. "The decline in plan numbers probably reflects the apprehensions – both external and internal – that developed after the fires of 1988," Jerry Williams, the national director of Fire and Aviation Management, acknowledged at a fire symposium in Missoula, Montana, in March 1993. "Certainly, 1988 demonstrated the kinds of risks that can accompany the wilderness fire program."[12]

Even as the fires burned in Yellowstone, there were a small number of scientists and conservationists who were brave enough to buck public sentiment and stand by national park and forest service decision-making. "Fire has a healthy role in the reproduction of the forest," said Michael Scott, regional director of the Wilderness Society. "We can watch the park be reborn. We are all midwives in this."[13] "When television and newspapers tell the world that fires 'are destroying Yellowstone', they are wrong," wrote M. Rupert Cutler in an op-ed in the *New York Times*. Cutler was president of Defender of Wildlife

16.2 About 32,000 elk, 2,700 bison, and 2,000 mule deer were roaming Yellowstone when the fires started. Fifty elk, four bison, several mule deer, five moose, and one grizzly bear died.

and a former assistant secretary of agriculture in the Carter administration. "The fires are not destroying Yellowstone, they're changing it," he wrote. "And in most cases, those changes will promote a greater diversity and abundance of plant and animals."[14]

Yellowstone demonstrated this. But it was a deadly fire in 1994, and the billion-dollar price tag that came with all the other fires that season, which convinced policy makers that the "business as usual" approach to managing wildland fire needed to be changed.

Fourteen firefighters died in July when a relatively tame fire began to burn in South Canyon on Storm Mountain in Colorado. Thirty-metre-high flames whirled out of control as they climbed upslope overwhelming the fleeing firefighters just metres before they could get to a safe zone at the peak. It was, as many in the wildfire community have noted, a low point in US fire-suppression history.

The following year an interagency consortium crafted a common federal policy for wildland fire. The budgetary ceiling for prescribed burns was raised. The emphasis on human safety was reaffirmed, but the risks that were inherent were accepted. The policy provided for mobile tactical support in years when the fire season was exceptionally severe.

Many of the terrible things that were supposed to have happened after the fires were finally doused by rain and snow didn't materialize. There was no long-term drop in tourism. Flooding did not occur because of increased runoff along forested slopes laid bare by the fire. Fish survived the carbon that washed through the rivers. Elk numbers remained stable over the long term as aspen shoots offered more food for them. (The re-introduction of wolves in 1995 and 1996 certainly helped keep elk numbers down.) Grizzly bears populations did not decline as some suggested; they simply altered their feeding habits for a while. Nor was there another big fire season because of all dead and downed trees.

The fire did more good than bad, as I saw on my hike in 2013 and others I took through Yellowstone. The park was not destroyed by those fires. With the help of transplanted Canadian wolves that found their niche in the Yellowstone ecosystem, mixed forests rose up to replace an old growth system that was increasingly vulnerable to disease and bark beetles. No longer were bears' summer food – thistle, biscuitroot, fireweed, grasses and sedges, dandelion, clover, spring-beauty, horsetails – being smothered. The root vegetables and berry bushes that bears feed on from September to November also came back once their seeds were exposed to sunlight.

The future of the park and others like it, however, are in doubt. Since 1950, snowfall has declined in the Greater Yellowstone Area by 53 per cent in January and 43 per cent in March. September snowfalls have almost disappeared. Annual snowfall has declined by nearly two feet since 1950. These and other factors setting the stage for fires to burn bigger, more intensely, and more often as temperatures continue to rise. If that happens, forests at low elevations will likely be converted to grasslands. Plants and animals that depend on this ecosystem could be threatened.[15]

17

BIG AND SMALL GRIZZLIES

Fire's role in creating better habitat was not embraced by many conservationists, who were increasingly worried about the indiscriminate clearcutting that was taking place in the 1980s and 1990s. They were rightly concerned that old growth forests, along with the caribou, spotted owls, and other animals that depend on them, were headed towards extinction. Even I was suspicious of the notion that a fire-scarred landscape can be better for some animals than a national park where hunting, logging, and oil and gas exploration are prohibited. But thinking began to change, influenced in part by a curious discovery by scientists studying grizzly bears in and around Jasper National Park and the Wilmore Wilderness of western Canada. Grizzly bears in the protected areas tended to be significantly smaller and thinner than bears that spent most of their time outside the national park.

I had firsthand insight into this discovery while accompanying biologists Gordon Stenhouse and Terry Larsen as they began searching for and capturing grizzly bears in the Jasper area in the late 1990s. One of the most telling moments occurred when we were helicoptering through a mountain pass. A light rain, mixed at times with wet snow, began to fall. The prospect of finding bears that day dimmed. Thinking out loud, the pilot considered turning back. But then the skies cleared, as they often do in between mountain snow squalls, and a female and a cub came into view. Seeing that she wasn't wearing one of his satellite-tracking collars, Stenhouse signalled his desire to capture the mature bear.

17.1 Grizzly bears and other animals do well when fire clears the forests, allowing berries and root vegetables to grow. Photo by Edward Struzik.

Following protocol, the pilot dropped Stenhouse and me off on a nearby mountaintop to lighten the load and remove the door on the passenger side of the helicopter. That gave him the steadiness he needed in the air to give Larsen an unobstructed and less shaky view to fire a tranquilizer dart into the female. It didn't take long for the helicopter to catch up with the bears. Larsen fired the tranquilizer into the female bear. She immediately slowed and fell belly-down on the ground as the cub hovered over her.

When the helicopter brought Stenhouse and me back to the scene a few minutes later, Larsen was keeping a wary eye on the tranquilized bear, knowing that she could suddenly rise from her slumber. The cub had run off. To make sure there were no surprises, Larsen and Stenhouse moved in with a rifle, calling out softly to make sure the bear didn't startle and wake up. As a precaution, the pilot kept the engine running. Satisfied that the bear was down and out, Stenhouse hooked her up to an oxygen tank to ease her laboured breathing. With only an hour to draw blood, extract a premolar to determine her age, secure some hair samples, and weigh and measure the length of the animal, time was tight and tensions were high as we waited for the veterinary team to be flown in from another nearby hilltop.

It didn't help when the cub, which was as big as its small mother, returned to the site, dancing around the perimeter before disappearing once again into a stand of spruce trees. Once Larsen curled back the lip of the tranquilized animal to see if she had a tattoo stamped from a previous capture, he discovered that she was the animal they caught in the first years of the long-term study. She was still as small as she was backing then, weighing only 230 pounds (104 kilograms). Otherwise, she looked to be in good health.

For the longest time, biologists had assumed that small bears like this were the product of natural selection. Given that winters are longer and harder there than they are on the west coast and in Yellowstone, it's tough for big bears to put on weight when they are hibernating for four or five months of the year and feeding mainly on roots, rodents, berries, and, occasionally, sheep, deer, and elk. There are no salmon runs on the east slopes of the Rockies as there are on the west coast of British Columbia and Alaska. Natural selection in this case favours the little guys, just as it does on the Arctic tundra.

But that is not necessarily the whole story, as University of Alberta scientist Scott Nielsen discovered when he began tracking and monitoring the fate of what would eventually total more than 100 grizzly bears in Stenhouse's study area. In 2013, Nielsen, Stenhouse, and others concluded that age and sex help determine whether a bear becomes big and fat. But they also discovered that habitat is just as important. Smaller bears, like the one we captured that day, tended to be found in colder, less productive environments such as Jasper, where a half-century of fire suppression has resulted in too many thick stands of trees and not enough roots and berries to sustain large animals. Disturbed environments, which included those burned by fire or logged by timber companies, offered up more food, albeit with the added risk of bears being poached or killed by vehicles in clearcuts.[1]

This finding was not altogether surprising to scientists who study grizzlies. Before the wildfires that burned so much of Yellowstone in the summer of 1988, biologists were monitoring the movement of thirty-eight bears that had radio transmitters attached to them. Twenty-one of them had home ranges in areas that got torched. Thirteen of them moved back into the burn areas immediately after the fire front had passed through.

In three cases, adult females with cubs stayed within a burn site while it was still smouldering. Some grizzly bears were seen feeding on carcasses of deer and elk killed in the fires. Others were seen grazing on newly emerged sedges and bluegrass and digging into logs and anthills for insects. As the years progressed, there was no evidence to suggest that the fires had affected the bears' movements or choice of den sites in any negative way.[2] In short, the bears benefited from the fires that had horrified so many people. The forest was not dying. It was giving birth to a new crop of roots, berries, and trees.

17.2 Bear scavenging in the afterburn of the Kenow Fire, Waterton Lakes National Park, 2017.

18

CLIMATE AND THE AGE OF MEGAFIRE

At 2 a.m. on the fourth of July in 1989, Bill McKnight, Canada's Defence minister, was roused from his sleep. It was a telephone call from Albert Driedger, the Government of Manitoba's minister of Emergency Services, asking him for military planes and equipment to help fight 224 forest fires in the province. "The situation has exploded on us," Driedger had told journalists the day before.[1] More than 24,000 people from thirty-two mining towns and First Nations communities were forced out of their homes in what was the largest evacuation in the province's history.[2] Only the city of Thompson and the town of Gillam were still populated.

Highways in the north were closed to everyone but firefighting teams, and Royal Canadian Mounted Police escorted convoys transporting food and fuel from Winnipeg to Thompson to help those who had fled smaller towns. An emergency system of water sprinklers was erected on a wooden railway bridge to prevent the fires from destroying the rail line that linked Churchill to Winnipeg. There was genuine fear that power would be lost across the province if the fires crossed roads and firebreaks and burned down transmission lines. More than two thirds of the province's electricity was being produced by hydroelectric generating stations on the Nelson River at the north end of the province.

Manitoba was subjected to its first "dark day at noon" for some time.

"It's like dusk in the middle of the day," said James Downey, the province's Northern Affairs minister. "You can't see the sun."[3] The situation was so dire that the Manitoba government was also asking the federal minister of Transport to waive rules limiting the number of hours a pilot could fly in a month. Most of the pilots fighting the fires were already at or near those limits. All available provincial civil servants were asked to report to their supervisors to relieve exhausted officials who had been working round the clock to co-ordinate the firefighting and evacuations.

It wasn't just Manitoba that was burning that summer. Some 1,200 people also had to be evacuated from the town of Sandy Lake in Saskatchewan. Four hundred people were forced to flee their homes in the northwestern community of Bearskin Lake. The Northwest Territories experienced twice as many fires (nearly 600) as it did on average. And British Columbia burned big for the first time since the devastating fires of 1938 and 1950 when the Chinchaga fire burned a wide swath through the northeast corner of the province into Alberta. In the first week of August, there were 2,600 forest fires burning in that province: 1,000 more than the year before. Fifty homes near the Okanagan town of Kelowna were emptied. Even cottage country in Muskoka, Ontario, was on high alert as several small fires in the region threatened to pick up steam.

This was the reason Manitoba could not call on other provinces to help. They were all too busy fighting their own fires. This was also the reason Manitoba joined the Great Lakes Forest Compact later that year. They really needed assistance from Ontario and neighbouring American states to fight fires as big as these. Although the number of fires and the area burned across the country appeared to have been declining since 1970, things changed in 1989. Data suggested a dramatic increase right across the country from 918,000 hectares burned in 1968 compared to the 7.6 million hectares that burned in 1989.[4]

Even so, in 1989 Manitoba was the standout. The number of fires that burned in May and June was two to four and a half times higher, respectively, than the twenty-year average.[5] Most everyone in the fire community had come to realize that something new and different was happening not just in Canada and the United States, but in countries like Australia, where bushfires in 1984–85 burned 3.5 million hectares and killed four people and 40,000 livestock just after an even deadlier fire – Ash Wednesday – killed forty-seven in 1983. Some 3,700 buildings were destroyed and 2,545 people were left homeless.

The surprisingly volatile behaviour of the Black Dragon and Yellowstone fires was still fresh in everyone's mind. Richard Rothermel, one of the world's leaders in assessing wildfire behaviour, said as much after he went to Yellowstone in 1988 to help park officials determine worst-case scenarios. Try as he and his colleagues did, they couldn't figure out what to plan for because, as he later explained, "the winds came again and again and the worst-case scenario happened almost weekly. It was an amazing season," said Rothermel. "Nobody had seen this combination of weather and fires before."[6] (Except, of course, the

Chinese in 1987, the Australians in the 1980s, and the lone fire ranger who saw what the Chinchaga Fire had done in 1950.)

The Black Dragon, Yellowstone, and northern Manitoba fires represented a turning point in North American fire history. Megafires – fires that burn more than 40,500 hectares – had been relatively rare up until then, partly because of aggressive fire suppression practices. Credit also goes to Smokey Bear, the establishment of North American Fire Compacts, the widespread adoption of the Canadian Forest Fire Danger Rating System, better training, and the recruitment of highly educated researchers like Mike Flannigan, who had graduate degrees in both meteorology and plant sciences when he joined the Canadian Forest Service in 1981.

After more than a half-century, Canada was also becoming a world leader in developing fire-fighting airplanes and helicopters with companies like Montreal-based Bombardier and Conair, a lesser-known British Columbia–based operation that had a fleet of twenty-eight helicopters and forty-six fixed-wing aircrafts in 1989. Unheralded as Conair was at the time, the company had planes and firefighting helicopters stationed under contract across Canada and in France, Spain, Portugal, and Australia. In August 1989, it was in negotiations to sell as many as five water bombers to Brazil.

So why, with all these planes, tools, and intellectual know-how in play, were megafires on the rise? Some suggest that poor funding for fire science was partly to blame. Money for research in the US Forest Service had dropped by 46 per cent between 1978 and 1990. There hadn't been any big fire-science initiatives in the US since Operation Firestop in 1954, which involved the military, the US Forest Service, and state and county forest organizations experimenting with new ways of supressing fires. The situation in Canada was just as bad, as economic imperatives in the energy and forestry industry overshadowed fire management concerns. Within the Canadian government, the Canadian Forest Service was constantly punted from one department to another in the 1970s and 1980s, never finding a minister charismatic enough to drive its agenda forward. Even its name was changed several times, as if anyone cared whether it was the called the Forest Branch or the Canadian Forest Service. It was even a government department for a time. By 1978, the number of full-time employees had dropped to 1,000 from 2,400 a decade earlier.[7] The situation improved after 1982, but not by much.

The outlier in all this was climate change. Two of the warmest years in recorded history occurred in 1981 and 1987. While the Yellowstone fires were burning in 1988, James Hansen, director of NASA Goddard Institute for Space Studies, was telling a Senate hearing that he was 90 per cent certain that the earth was warmer than ever recorded and that the cause was rapidly rising greenhouse gas emissions. He predicted that the likelihood of freak weather, the kind that triggers prolonged drought, intense fires, tornados, and hurricanes, would increase.

It was a time when government scientists didn't have to worry about saying things that politicians did not like to hear. "If climate change is coming, this sure looks like it," said Charlie Van Wagner, when reporters asked him what he thought about the 1989 fire season in Canada. Van Wagner suggested that if the warming trend continued, fire management costs would be stretched to the limit.[8] It was an astute prediction that Brian Stocks and others would hammer homes multiple times in the coming decades.

Don MacIver, a meteorologist with Canada's Atmospheric Environment Services, summed up the situation succinctly when he pointed out that the area burned in 1988 was higher than anything seen in recorded history. "The rising fire damage has become a major scientific and forestry issue," he said. "There's no question about that at all." At time he was working on the first Intergovernmental Panel on Climate Change, which was established in 1988 by the United Nations Environment Programme and the World Meteorological Organization. The concern was no longer the cooling that might take place in the event of a nuclear war. It was how forests, grasslands, oceans, and lakes would respond to steadily rising temperatures.

The IPCC report of 1990 did not speculate much on the potential for wildfires burning bigger, faster, and hotter as they did in China in 1987, Yellowstone in 1988, and northern Manitoba and many parts of Canada in 1989. But James Harington, a scientist at the Canadian Forest Service in Petawawa, was already setting the stage for these foci of interest in 1986 when he submitted a paper on the subject to the *Canadian Journal of Forest Research*. Harrington suggested that fears about a nuclear winter are "not unfounded." But the purpose of his review, he stated, "is to alert readers to the reality of climate change and to provide them with information on which reasonable action can be based." If the Earth warmed by 2.5°C by the year 2050, as many modellers at the time predicted, the impact on the boreal forest of Canada would be profound, according to Harrington. The past, he added, "is not a sure guide to the future (because of greenhouse gas emissions) … The implications for forestry in Canada are immense and therefore climate change should become a vital aspect of planning and policy."[9]

Mike Flannigan had joined the Canadian Forest Service a few years earlier, desperate to escape the grinding shift work that he had to endure while working as a meteorologist with Environment Canada. He loved weather forecasting, but with a new child in the household, he needed a job with more stable hours. A PhD in plant sciences made him a shoo-in for the job in forestry when he applied. Harrington was Flannigan's supervisor at the Forest Service. He encouraged his protégé to pursue the climate change angle. Flannigan, however, was sitting on the fence at the time, not really sure whether humans were responsible for what was clearly a warming world. His heart wasn't entirely in it when he was asked to speak on behalf of his supervisor in 1985. It was the first of hundreds of talks he would make on the subject.

But then several things clicked as Flannigan began looking into climate change more deeply following the Manitoba fires. There had to be some reason why the amount of area burned in Canada had more than doubled since 1970, as Van Wagner had concluded two years earlier.[10] And with the Manitoba fires coming just a year after Yellowstone and two years after the Black Dragon Fire, new thresholds were being set. With Van Wagner's experience and deep background, he and Flannigan published a paper in 1991 that suggested a 46 per cent increase in fire severity with a doubling of greenhouse gas emissions. These increases would not necessarily wreak havoc on Canadian forests," they stated cautiously. But they implied that they might.[11]

Brian Stocks and his colleague Michael Weber were much more forthright on the subject of the future in a report they collaborated on three years later. "It was obvious that a warming climate in the future would require a rethinking of fire management priorities and a re-evaluation of fire in forest management in North America," they wrote. They warned that there would be no "quick fix solutions." Increased fire activity would "strain current levels of fire suppression resources, but may also adversely affect boreal forest distribution with a concomitant reduction in plant and animal biodiversity."[12]

Until that time, Stocks had behaved as any pure scientist would, conducting experimental burns in the field, writing up what he discovered in scientific journals, and developing new and improved iterations of fire danger rating and fire behaviour prediction systems. But as budget cuts and political indifference continued to impede meaningful reforms in fire management, Stocks decided that nothing was going to get done if he and others just sat back in their labs and cashed their pay cheques until it came time to retire.

At some point, he realized, climate warming was going to intersect with the wildland-urban interface, that line where the forests and dry brush intersect with houses and businesses. "By the late 1980s, we knew that growing Wildland Urban Interface problems in Europe, Australia, and the United States were coming to Canada," he said in an interview with me. "And because of our involvement in many international, cross-disciplinary associations, we knew that climate change would exacerbate that challenge. There were just too many people moving into or working in the boreal forest at a time when temperatures were warming and drying those forests. Something bad was going to happen, sooner rather than later, we thought."

19

THE HOLY SHIT FIRE

In the summer of 2003, I was flying across Kootenay National Park with Rick Kubian and Rob Walker, two of Parks Canada's more seasoned wildfire managers. It was a cool, wet day and the leaden clouds floating below us were marbled with the smoke that was spiraling lazily upwards from the charred, smouldering mountain landscape below. It had been forty days since the fire started. Both men were spent, but sleeping and breathing more easily for a change. Nothing in the weather forecast suggested the fire would come back to life.

It was then that I finally asked the question I had been reluctant to ask until this time. Why was everyone fighting these fires in the Rocky Mountain parks referring to this one as "the Holy Shit Fire"? I could see that Kubian, who was up front in the helicopter, and Walker, sitting beside me, were both smiling.

"Every time we put someone in a helicopter to have a look at this fire, they'd say, 'Holy shit!'" Kubian replied. "Every time I flew over it, I would get a lump in my throat looking over the near continuous stand of old-growth forest between Lake Louise and Banff just waiting to go up in flames." Then Walker added, "Only a handful of people know how close this fire came to lighting up the Banff side of Kootenay. At times, the fire had its own mind, doing whatever it wanted to do. Fortunately, we made the right decisions and contained it. But it was awful close. We learned a lot of lessons."

The 2003 fire season was notable because it started in another part of the world. The Australian Alps experienced their largest bushfires in over sixty years, with an estimated 1.73 million hectares consumed. Bushfires burnt across

Victoria, New South Wales, and the Australian Capital Territory during that country's worst drought in one hundred years.[1] Canadian water bombers were brought to Portugal to help that country deal with its worst fire season ever.[2] Several people died in the French Riviera that summer as forest fires sizzled along the Mediterranean Sea.

In North America, the fires were notable not for the area burned but because so many cities, towns, national parks, industries, and historic sites were in harm's way. This was a harbinger of the kind of fire season that James Hansen, James Harrington, Brian Stocks, Mike Flannigan, and others had been predicting almost two decades earlier, and what Stephen J. Colombo, Robert S. McAlpine, Mike Flannigan, and many others had been forecasting for Ontario more recently.[3]

For a good part of that summer, more than 2,000 people were forced to evacuate the Crowsnest Pass region in southwestern Alberta. Nearby to the east, Waterton Lakes National Park was on standby as 13 per cent of neighbouring Glacier National Park in Montana was being burned by twenty-four fires, some of which spread so rapidly that diners at Lake McDonald Lodge left their meals half-eaten as they were ordered to evacuate. Two firefighters were killed while attempting to suppress a lightning-triggered fire in Salmon-Challis National Forest in Idaho. Oregon fared a little better, but not by much. Some 1,000 people had to be evacuated when two fires in the Central Cascades joined together. At one point, there were 2,379 firefighters on scene supported by 112 pumper trucks and fourteen helicopters. The cost of supressing the fire totalled $38 million. Touring the area, then president George Bush had a difficult time describing what he saw as flames flew from treetop to treetop. "It's the Holocaust; it's devastating," he said after the rocky helicopter ride. "We need to thin our forest in America. Our citizens must understand there are millions of acres of forests around this country that are vulnerable to catastrophic fires because brush and small trees have been collecting for decades."[4]

In Jasper National Park, Parks Canada wildfire managers were trying to do just that when a prescribed fire in Jasper got out of control. It might have torched a path to the town of Hinton twenty kilometres away had it not been for Parks Canada wildfire managers dropping fuel on the forest to get the incoming fire to turn on itself. Some of those managers had just returned from Quebec, where a fire in La Mauricie National Park was getting a lot of attention because it was in Prime Minister Jean Chretien's riding. To the south of Jasper, Banff was being threatened on two fronts at the height of the tourist season. The first came from an out-of-control prescribed burn that ignited east of the townsite near Canmore; the second from that enormous fire in Kootenay that was chugging along like a slow-moving freight train on both sides of the mountain highway that runs through the park between Revelstoke and Banff.

Wood Buffalo, Mt Revelstoke, and Prince Albert national parks also burned in the weeks before the interior of British Columbia lit up around towns and cities such as Kamloops and Kelowna. At one point, the military was brought in to fight those fires. Triaging became the order of the day. Only fires that threatened people were being managed. More than 40,000 people were forced to leave their homes. And then, as if to put an exclamation point on all that had happened that year, California lit up like it had never done before when three fires burned through San Diego county. The Cedar, Paradise, and Otay fires started on 25 and 26 October, as powerful Santa Anna winds blew in from the mountains towards populated areas. Sixteen people lost their lives; 3,241 structures were destroyed. Suppression costs topped $43 million. The Cedar Fire alone, at 110,578 hectares, was the largest in California history up until that time.[5] Not since the Laguna Fire of 1970, when eight people died and 71,000 hectares burned, had the state seen anything quite as intense and destructive.

Looking back on the 2003 fire season, almost everyone agrees that it wasn't so much a matter of why it happened, but why it didn't happen sooner. Warning signs had been flaring up periodically since Yellowstone burned in 1988. Until then, there had been only five megafires in the United States in the seventy-eight years since the Big Burn of 1910. In the fifteen years that followed Yellowstone, there were twice that many.[6]

In Canada, new records were being set since Manitoba burned in 1989. In 2001, the Chisholm Fire in Alberta established a new threshold for intensity, replacing the ominous record set by the Vega Fire that ripped through the same region in 1968.[7] For the second time, Alberta's fire season started a month early. The only other time that happened was the previous year.

It was not just a provincial trend, but a North American one.

Three fire seasons stood out among the fires that burned in between Yellowstone in 1988 and the 2003 fire season. The 1991 Tunnel Fire that killed twenty-five people and injured 150 others when it tore through Oakland was one of the shockers. The South Canyon Fire that resulted in the deaths of fourteen wildland firefighters in 1994 was another. That was the year California used gang members looking to restart their lives as a way of helping volunteer and professional firefighters solve an increasingly challenging fire problem in the state.[8] The fire that really signalled a breakdown in the system was a prescribed burn that got out of control and sent 18,000 people fleeing the towns of Los Alamos and White Rock in New Mexico before destroying 280 homes and threatening nuclear weapons storehouses at the Los Alamos National Laboratory. A headline in the men's magazine *Maxim* summed it up for anyone who might not have understood the significance of this Cerro Grande Fire. "An out-of-control wildfire. A nuke factory with enough plutonium to wipe out the entire Southwest. A handful of exhausted firefighters. Just how close did we come to annihilation?"[9]

The "Holy Shit" fire in Kootenay was one of those "what if" fires. No one died. No buildings of significance were destroyed. The fire, however, was one that further underscored the fact that the "business as usual" approach to managing fires was fraught with danger in a world that was warming even faster than climate scientists had predicted.[10] The first warning came in April in a memo from Mark Heathcott, the national parks fire management coordinator for western Canada, obtaining and deploying resources such firefighters, helicopters and water bombers. Heathcott was in the habit of forecasting the upcoming fire season on April Fool's Day, using the handle "Charlie Lima" (aka "Chicken Little") when the forecast was scary. He sent out a map that spring illustrating how most of Canada, and especially the west, had gone through one of the driest falls and winters on record. The meteorological data made his forecasts sound ominous. So did those that were being issued by other agencies on both sides of the border. All were expecting the 2003 fire season to be one of the most challenging yet. "El Niño," Heathcott added in his memo, "was going to hang in there, until at least the end of summer." "Look for an active fire season in Mt. Revelstoke-Glacier, Kootenay and Yoho," he said before signing off.[11]

The second warning came when spring weather produced almost no snow or rain in advance of a summer that turned out to be one of the hottest and driest on record in both southwestern Canada and the northwestern region of the United States. El Niño events tend to influence climate and weather patterns in this part of the world, but no one had seen anything quite like the baking that took place in the western forests in June, July, and August. In fact, heading into the summer of 2003, southern British Columbia, Montana, Idaho, parts of Oregon, and Washington had gone through their driest three-year period on record.

At the height of the drought, thirty-two massive dust storms swept across the prairies. Stream flows in many places were between 10 and 20 per cent of normal. Water flow along the mighty Fraser River in British Columbia was at a historic low. Salmon were dying en masse in lethally warm streams. Adding to the woes of farmers and rural dwellers were swarms of grasshoppers that chewed up withering crops, and in some cases, the paint on houses. In one Saskatchewan district, there were as many as one hundred grasshoppers per square metre of land.[12] And if all this wasn't enough to prove that climate change was raising temperatures to new thresholds, the mountain pine beetle population exploded as it began a march westward that would eventually cover more than eighteen million hectares of forest – an area five times the size of Vancouver Island.[13] Winters just weren't cold enough to knock them down.

Heathcott remembers the 2003 wildfire season for another important reason. When resources were virtually almost tapped, he was asked to send help into polar bear country to prevent a fire from burning the Prince of Wales historic site near the west coast of Hudson Bay. The building survived, but several polar bear dens in the boreal forest of Wapusk National Park were destroyed.

Typically, Heathcott would have been able get help from the provinces and neighbouring states to deal with the unusual situation, but the competition for resources was fierce that summer. Supply was so tight all across North America that one helicopter – an aging Hughs 500 that he managed to acquire from the United States – broke down twice on the way to the Kootenay fire. When it finally arrived, it had a pilot who also appeared to have seen better days.

Heathcott and his colleagues were lucky because Nik Lopoukhine, the director general of Parks Canada, was a forester and he understood fire. He signed off on an emergency response plan when Heathcott warned him in the spring that there was likely to be an intense fire season. Lopoukhine's directive compelled national park superintendents from across Canada to send employees to help suppress fires that might get out of control.

Even then, the Kootenay fires proved to be too intense to handle in traditional ways, as blazingly hot temperatures and low humidity sapped moisture from every living thing. The only saving grace was that the park typically doesn't see the number of tourists and backcountry hikers that Glacier officials had to round up and evacuate that summer. To this day, Rob Walker isn't so sure that the helicopters and ground crews that were deployed in Jasper and elsewhere could have contained the situation in Kootenay.

Walker and his colleagues were seeing fires do things that summer that he and his colleagues had never observed or even heard of before. Notable among them was what happened when firefighter Patrick Langan had a helicopter drop a bucket of water on a fire that was barely smouldering along the shore of Taylor Lake, halfway between Lake Louise and the Castle Junction. Instead of putting the fire out, the downdraft that came with the water falling flared things back up. In ten seconds, it went from a no-flame situation to one of huge candling and torching. "Candling" is the firefighter's word for flames that go from bottom to the top of a tree or a clump of trees. With more than 300 firefighting missions under his belt, Langan said he had never seen such "outrageous fire behaviour."[14]

The behaviour of these fires called for extraordinary measures and unorthodox decision-making. At Taylor Lake, Langan made the unusual decision to call in an air tanker. This was an extremely costly move that necessitated the accompaniment of a "bird dog" airplane, a small one that spots the fire and leads water bombers to them. But Langan felt it was necessary to prevent fire from spreading southeast, where it had the potential to overrun the Banff townsite and the 25,000 tourists who were visiting there. In the meantime, Walker, Kubian, and Jeff Weir, a fire specialist from Prince Albert who brought in sprinklers to help contain the fire, decided that given its intensity, they had to stop it from chugging past Castle Junction into the old growth forest in Banff. If it got

19.1 The Vermilion River Fire of 2003 in Kootenay came close to jumping into Banff where it would have likely run to the townsite. Photo by John Niddrie.

beyond that line, they realized, a backfire would have to be lit to get the fire to turn on itself. "This was not a normal fire," Walker recalled. "It had a mind of its own. It defied everything that one learns about wildfire behavior. I had expected that the deciduous vegetation along the slide paths on the mountain slopes would slow it down because deciduous trees hold more moisture and burn more slowly than the conifers. But those trees just vaporized in the fire's path. It was like a dragon breathing fire on dry grass."[15]

The day of reckoning came during the third week of August, when the fire took another spectacular run towards Banff. A meeting was called for all senior park managers. There was no office to meet in, just the tailgate of a pick-up truck on which maps were laid out. The decision was made to light a backfire, even though the winds were stronger than they should have been based on fire behaviour computer models that were being used. "I'll never forget what happened next," Walker recalled, when I interviewed him about what had happened

that day. "It was like a tailgate party. Michel Boivin (the superintendent of Lake Louise) was on hand, and we came over and gave him the bad news. We told him that we were better off risking failure than letting this thing come to us. You could tell he was nervous."

Boivin understood that he would be damned if he said no and allowed the fire to spread into Banff, and damned if he gave the OK and the backfire failed to stop the fire's run. Shortly after he threw caution to the wind and called for the backburn to be ignited, the "freight train" crashed into the fireguard. From the bird's eye view offered by a helicopter, Walker and Kubian could see the giant leading plume of smoke standing up on itself and coming to a halt as it ran out of fuel to burn.

There were several reasons, apart from El Niño and resource shortages, that made this fire and the others that burned in 2003 harbingers for the future. Climate warming was clearly intensifying droughts, raising temperatures, speeding up wind, increasing lightning strikes, and cheering on insects, like the mountain pine spruce beetle, that favoured old trees because they did not have strong enough defence mechanisms to ward off an attack.

Fires in rapidly warning forests were also intersecting dangerously with the increasing number of people living, working, and playing in those forests.

19.2 The spruce beetle, like the mountain pine beetle, has ravaged many aging forests throughout Canada and the United States. Photo by Edward Struzik.

Between 1990 and 2010, one in three houses built in the United States was in the wildland-urban interface.[16] The same kind of developments were occurring in Canada. Throughout the 1990s and into the twenty-first century, the city of Kelowna had one of the fast-growing populations in the country, increasing at a rate of 4.5 per cent annually.[17] Many of the people who were living, working, hunting, and camping in this interface were responsible for the big fires putting them and others in harm's way. The Cedar Fire in California was a perfect example, having been started by a novice hunter who got lost in the woods and lit a fire to attract attention to his whereabouts.

It's impossible to say that the 2003 fire season was directly linked to climate change. But it fit nicely into the models that Mike Flannigan and others had been working on. This time, it wasn't just First Nations and mining communities that were at risk, as it so often had been in the past. It was wealthy people, not unlike those in Bar Harbor in 1947, who could afford to build big expensive homes in the forested areas around San Diego country and the Okanagan. Because many people were building homes with cedar shakes and wood siding and decks, and landscaping them with highly flammable ornamental plants like cedar and juniper, there was no stopping a wind-driven fire shooting embers five kilometres on a roasting hot day.

Fire managers are often shown no mercy when fires like these force evacuations, kill people, and damage homes and businesses. Inquiries inevitably follow and heads sometimes roll, as they did after the Salmon-Challis Fire, and politicians who want to turn the clock back to the days when every fire was fought diligently, dutifully dole out money to compensate the victims who should have seen it coming. Not a thing had changed since John Irwin wrote in 1947 that "the general approach by governments has been to spend as little money as possible in advance, but when fire comes, to parade an all-out determination to fight for the people's interest, regardless of cost."[18]

Most of the people who moved into the wildland-urban interface didn't see the fires coming because urban planning, building codes, zoning regulations, fuel management, and public education were not addressing the issue in any meaningful way, despite the recommendations of the inquiry into the Laguna Fire of 1970. Ronald Reagan was California's governor back then and, like other governors before him, he embraced the laissez fare attitudes of the nineteenth and twentieth centuries. When Arnold Schwarzenegger won the election a month after the Cedar Fire was extinguished, the tradition continued. To these politicians, fire was like an earthquake, flood, or mudslide that occurs and then is gone. It was left to the insurance companies and maybe the state legislatures to compensate people for their losses.

And nothing of consequence happened in Canada when former Manitoba premier Filmon was appointed to head an inquiry into the Okanagan fires. The expert panel that Filmon chaired laid out a blueprint for fire management in as comprehensive a manner as General Howard Kennedy did with his Report of the Royal Commission on Forestry in 1947. As Filmon himself stated, and almost pleaded, the inquiry he led was a "once in a lifetime opportunity" to get on top of the increasingly dangerous trend of megafires.[19] But after promising to follow Filmon's suggestions, British Columbia passed on the opportunity. Two years after the report was released, the auditor-general produced a scathing indictment, noting that only seventeen of the seventy-four recommendations made by the expert review committee had been implemented.

The Canadian Council of Forest Ministers promised to deal with this oversight in 2005, when it produced its own plan for a national wildfire strategy months after Mike Flannigan, Brian Stocks, and their colleagues published yet another study which predicted a 74 to 118 per cent increase in area burned by the end of the century if greenhouse gas emissions rose as predicted.[20] But that plan also stalled as the Canadian Forest Service moved away from wildland fire to focus more on issues related to insects such as the mountain pine beetle. Stocks and Flannigan were mortified by these developments, so much so that Flannigan began considering the possibility of a university appointment if one happened to come up. Stocks, who was working part-time at that point, decided to pack it in.

In the years that followed, nothing happened to make him regret his decision. From 2006 to 2010, the Canadian Forest Service's wildfire research budget was a paltry $1.3 million, almost ten times less than the money that was going to insect-related issues. Less than 1 per cent of the money that was needed to create the national wildfire strategy was forthcoming.

This was not entirely surprising, given that Canadian Forest Service was part of Natural Resources Canada, which was also responsible for energy. And that, in large part, explained the discrepancy in funding and the increasingly inability of the provinces to get on top of the wildfire problem. Energy, not forestry or agriculture, was now king of the economy. In 2000, the forestry industry outperformed the auto and oil and gas industries as the single largest contributor to Canada's trade surplus. A decade later, giant corporations such as Abitibi-Price, MacMillan Bloedel, and Noranda Forest that had once employed more than 300,000 Canadians making newsprint, tissues, fine paper, cardboard, floor joists, and wall studs were gone. More than 100,000 jobs had been lost.

Stocks recalls vividly meetings in which there was "a fair amount of yelling and screaming ... We had some scientists who were highly motivated, who believed strongly in what we're doing to the earth and the atmosphere, and then policy folks who believed in the economy." The energy policy folks prevailed, helped by rock-bottom royalties, generous subsidies, and bargain-basement prices for oil and gas leases. These incentives, combined with the rising price

19.3 In 2011, the entirety of Slave Lake, Alberta, was evacuated. A third of the buildings were destroyed and there was $700 million in damage.

of oil and gas, spurred on what was already a phenomenal run on oil sands developments, fracking, pipeline construction, and all the roads and seismic lines that were needed to support them. By 2010, 34,773 wells, 66,489 kilometres of seismic lines, 11,591 kilometres of pipelines, and 12,283 kilometres of roads had been built in west, central, and northern Alberta. That didn't include the vast areas of forest being logged.

As Alberta and northern Saskatchewan became saturated with energy developments, the wilderness of northeastern British Columbia, where the Chinchaga Fire ignited in 1950, became the new frontier. Sales of Crown petroleum and natural gas rights in the region reached a record $2.7 billion in 2008, followed by a combined total of almost $1.7 billion in 2009 and 2010. Over 90 per cent of these investments were directed toward the exploration and development of British Columbia's shale gas regions.[21]

As these energy developments expanded, people followed. From 1972 and 2007, western Canada attracted 629,000 more people than it lost, while the rest of Canada combined suffered a net loss.[22] Most of these people ended up working in a boreal forest where spruce, pine, and aspen had been allowed to burn until the Chinchaga Fire of 1950, and less frequently as the years passed. By far, most of the area burned in North America burns in the west.

Once these newcomers settled in, it was increasingly impossible to let nature take its course as climate change – spurred on by those greenhouse gas emissions from heavy oil, fracking, and more traditional oil and gas and pipeline developments – set the stage for even more fires. Tried and true strategies such as "let-burn" and "prescribed burns" became increasingly difficult to allow. Fire suppression costs rose, resulting in less money for fire science. There would be no experimental burning programs in Canada after 2000.

When a third of the town of Slave Lake in Alberta burned in 2011 after forcing all 700 residents out, everyone in the fire world knew that there was no turning back the clock. If "business as usual" was going to prevail, something big and terrible would happen, as it did eventually in the oilsands town of Fort McMurray in the spring of 2016. Climate, and the city that was warning it with fossil fuel emissions, would demonstrate how broken the system of managing fire on the landscape had become.

20

THE PYROCENE

Three days after a wildfire ignited eight kilometres southwest of Fort McMurray on 1 May 2016, Lorna Dicks, the acting superintendent of the RCMP detachment, was in a makeshift war room working around the clock with colleagues planning what to do if the fire made a run towards the city. In the absence of a clearly defined strategy from the regional emergency response team, she realized that the Mounties needed to get up to speed. In the middle of that afternoon, Dicks was on a conference call with RCMP officials from across the country when she got a text message from her sister-in-law, along with a photo of a big fire burning behind a row of houses.

"What do you think I should do? Should I get Sam out of school?" her sister in-law asked her. Puzzled, Dicks asked her where she took the picture. "From the front steps of the house," was the response.

Dicks was flabbergasted, because no one had told her the fire had entered town. She had been given the impression that a soggy peatland that stood between the southside neighbourhoods and the forest would slow or stop it. Neither she nor the firefighters knew that the vegetation maps they were using to determine how the fire would move were hopelessly out of date. The fen had been drained to grow trees in an experimental forestry project conducted many years earlier. Instead of slowing the fire, the now-forested fen fanned the flames.

"I told her to get Sam – Samantha's my niece – and get out of town right away," Dicks said. Minutes later, Inspector Mark Hancock, her colleague and coleader, got a similar message with photos of the subdivision of Abasand

20.1 The Horse River Fire in 2016 forced the evacuation of 88,000 people.

burning. At that point, Dicks looked out the window and saw a normally quiet street rapidly filling with cars heading to the only highway out of town. "Clearly something was happening, and it was happening fast."

Dicks told me this story in that war room a few months after 88,000 people were evacuated in what was then the costliest natural disaster in Canadian history. "From that point on, we went into full-court press. It was chaotic. We went from going zero miles per hour to one million miles per hour. I made a call to district headquarters for mandatory overtime and got it right away. We deployed almost all our 135 police officers, making exceptions for married couples who had kids. One of the parents was allowed to make sure the kids were safe. We had our officers on the street directing traffic and going door to door. In the next five days, we had to move our emergency operations center four times. It was like a game of battleship. The fire made a move, and we responded by moving. In ninety-six hours, I got nine hours of sleep. All the guys on the team did the same."

The Horse River Fire was the catastrophic one that Canadian fire scientists like Mike Flannigan, Mike Wotton, and Brian Stocks had been predicting for years. Slave Lake, which burned badly in 2011 and forced the evacuation of 7,000 people, was just a rehearsal for this tragedy in the boreal forest that ended a baseball throw from the edge of town. The big problem was that few people in the oil sands capital had learned from what happened at Slave Lake.

The fire began on a hot and dry Sunday. Helicopter pilot Heather Pelley remembers it well because she was on standby with a small firefighting crew at the Grayling Creek Fire Base, thirty-five kilometres south of Fort McMurray in northern Alberta. Shortly after 4 p.m. that day, Pelley was playing guitar on the patio when a "dispatch alert" came in with a notice that smoke had been detected in the region. Once a call like that comes over the radio or satellite phone, a pilot is expected to be in the air with a four-person helitack (HAC for short) in ten minutes.

Pelley and Dave Mulock, the helicopter pilot who was already on the scene, had seen this coming when the Alberta government had deployed them and contract firefighting personnel across the province a week earlier than usual. The relatively mild winter had produced almost no snow. Spring followed with temperatures that were almost twice the average. With dry grass, brittle shrubs, and a never-ending forest of spruce, pine, poplar, and aspen trees thirsty for rain, everyone expected it to be an early fire season. More than half the fires that had burned since 1994 ignited that month. This was already the ninth fire in the Fort McMurray region that spring, and the 329th to have ignited across the province. Many more were burning to the west in British Columbia and to the east in Saskatchewan.

As hot, dry, and extremely windy as it was that day, Pelley still expected this call to end up being the "one-tree wonder" that initial attack crews often deal with when lightning, a campfire, an arsonist, or the scorching hot exhaust pipes of an all-terrain vehicle (ATV) ignites a fire in the forest. But when the helicopter cleared the trees that surround the base at the Grayling Creek station, she realized how wrong she had been. "I thought, 'Holy shit!' How long has this fire been burning?' It was fifty kilometres away and we could see this big plume of white smoke in the distance. There was no need for the dispatcher to give us the coordinates, as they usually do for an initial attack, so we pointed the nose towards it and off we went," she said.

The Bell 407 that Pelley flew is one of the fastest, most maneuverable helicopters on the planet, easily capable of reaching speeds of 130 knots (150 miles per hour). In the short time it took Pelley and the HAC crew to get to the fire with hoses, hand tools, backpack pumps, and a bucket capable of scooping up to 180 gallons of water, the fire was candling, bending with the wind, billowing black smoke over a high-power transmission line, and pushing west and southwest with winds gusting up to thirty-three kilometres per hour.

Mulock had already dropped buckets of water on the fire before putting Helitack crew leader Sebastien Rioux and his team on the ground. Pelley had been involved in sixty fires the year before, and she could already tell that there was nothing they could do to stop this one. A call had been made to bring in an air tanker based 250 kilometres to the south outside the town of Lac La Biche.

All she and Mulock could do was to continue dropping buckets of water to keep the HAC teams out of harm's way. "It was crazy," said Pelley. "We had to hover high above the smoke and the powerlines, struggling against gusty winds. Before any of the water got anywhere near the ground, it evaporated."

The Electra air tanker that was on its way from Lac La Biche carries 1,750 gallons of chemical fire retardant that can put out a small fire or slow the momentum of a larger one long enough for ground crews to get on top of it. Traditionally, it is guided in by a lead aircraft, or "bird dog," like a Turbo Commander 690. The bird dog flies twice as fast as Pelley's helicopter. There were two people on board this flight: a specially trained pilot and a government-employed air attack officer who decides when and where the retardant is to be dropped. Once a bird dog arrives at a fire, the air attack officer takes charge of everything, including the helicopter pilots. "The guy in the bird dog was super calm," recalled Pelly. "But you could tell by the urgency of his radio voice that he was really concerned. Right away, he sent out word out that this was a priority fire."

With the benefit of hindsight, it would have been better if a tanker had been based in Fort McMurray that month. But that's a justifiable oversight when resources are too thin to cover the entire province. It's even harder to second-guess the decision to divert that air tanker away from the Horse River Fire in order to deal with another one reported almost simultaneously. This one was inside the boundary of Fort McMurray, where it posed a more immediate risk. The fire was also smaller and more likely to be extinguished.

The decision to divert the air tanker to Fort McMurray proved to be correct, as an expert review committee, which included Brian Stocks, concluded when it assessed what went right and what went wrong. Reading their report as well as another is an agonizing exercise because the terms of reference in one case were too narrow to deal with all that needed to be addressed, and reviewers in both cases tried to be diplomatic. Little was said of why the fire got into town without notice, or why the emergency response team waited so long to mobilize the evacuation. Unlike Americans (who tend to drop the hammer when things go sideways, as they did when the Salmon-Challis Fire resulted in the deaths of firefighters), Canadians tend to be polite. These reports were no exception. "Good decisions and hard work are not always rewarded with good outcomes given the complexity of wildfires in these conditions," was how one of the reports put it. A lot of good decisions were made. But there were also many awfully bad ones.

The decision not to close the forests in the Fort McMurray region to all-terrain vehicles was one of them. ATVs cause fires; a 2002 study by the Alberta government concluded that at least eighty-three wildfires had been started by ATVs in the previous twelve years. The study's authors suggested that spark arrestors and heat-dispersion shrouds could be added to ATVs to reduce the

20.2 Wind carrying embers two kilometres from the fire front burned down many buildings.

possibility of ignition by keeping the fine fuels on the ground out of contact with the vehicles' exhaust system. That did not happen.

Closing the forests to ATVs would have made even more sense because the relative humidity that Sunday was as low as 15 per cent. The wind was gusting at speeds of thirty-three kilometres per hour. The fuel on the ground had a moisture rating of 4 per cent. As Mike Flannigan later told me in an interview: "Theoretically, anything (a spark or lightning strike) that looked at this forest sideways would have resulted in a fire." The forests, however, remained open. And although the official cause of the fire was never determined, it's now believed that an ATV was responsible.

Once the fire was underway and grew more complicated by the hour and the day, the Incident Command System deteriorated, with Alberta Forestry and the Municipality of Wood Buffalo often operating independently of one another and, in some cases, viewing each other with suspicion. The municipality, according to a review done for the city, may have not strictly adhered to predetermined shifts within its incident command system, "resulting in some confusion over who was 'in charge' at any one point in time, and who had the ultimate authority to make critical decisions."[1]

Inexplicably, members of the Regional Emergency Response Centre sometimes did not understand the arcane jargon used to describe the progression of the fire. Some used personal cell phones instead of the phones that were assigned to them. It was such a gong show that Slave Lake fire chief Jamie Coutts was told to go home when he and two colleagues arrived and offered to help. They had first-hand experience that few of the firefighters in Fort McMurray possessed. Had the Fort McMurray fire chief not arrived in time to listen to Coutts's appeal to stay, he would not have witnessed what Coutts later told me was an inexcusably chaotic response to the fires that were burning or threatening the city.

The inexperience of so many in dealing with a complicated fire like this resulted in a lot of second-guessing and complaining about the data, or the lack of data, that was coming in. The fact that the firefighters were constantly being forced to retreat as a result of the shifting winds and intense nature of the fire didn't help. Their reluctance to light backfires was also an issue. Alberta Forestry officials decided against the practice, fearing that a shift in wind would send fire racing towards Fort McMurray. Nonetheless, one of the review committees believed that these tactics might have been employed earlier with some success. This lack of unity and the absence of a clear line of command was one of the reasons Suncor and Syncrude, two of the biggest giants in the oilsands industry, and Enbridge, the pipeline company, ended up hiring their own experts and pilots to protect their assets.

The biggest mistake was when the regional emergency response team decided to wait for the fire to make a big run to Fort McMurray before ordering a general evacuation. On Sunday evening, the fire was less than eight kilometres away, a distance that many fires in the past had covered easily in an hour. There was no evidence to suggest that this one would act any differently, give the forecast for the following day. By 11 a.m. on 2 May, it had grown to 1,285 hectares. Lighter winds and cooler temperatures that morning slowed its momentum. But as humidity levels plummeted and temperatures rose to 27 degrees Centigrade in the afternoon, the fire moved rapidly westward, reaching the Athabasca River by 8 p.m. and then shooting embers one kilometre across the river overnight. If there was a time to get everyone in their cars and on buses, it was probably then. Pelley told me that when she was in the air that night and saw how the fire was behaving, she would have liked to put a megaphone on her helicopter struts and tell everyone in town to get out.

It was hard to fathom what Darby Allen and his colleagues were thinking when, at 11 a.m. the following morning, they held a news conference warning people that fire conditions were extreme and that residents should be prepared to "act on short notice." To then suggest that people to "get on with their lives and take their kids to school" confused rather than reassured them.

Like almost everyone in Fort McMurray, Adrien Welsh had no idea that the fire would arrive just two hours later, even as thick smoke started drifting into town. All she had to go by was what she saw on Facebook and heard from word of mouth. There was really no system in place to tell people to get out, which was remarkable given all that had been learned from Slave Lake in 2011 and the Okanagan fires of 2003. "In Central Arkansas where I'm from we have tornados," she told me a few months after she returned to her home in Fort McMurray. "In northern Alberta we have wildfire and lots of smoke. You get used to it. Nothing I heard told me any differently." Her calm turned to genuine concern when that thick smoke turned day into night. "I took a picture and sent it to Lucas, my husband, who is a firefighter for one of the oilsands companies." she said. "He told me to start spraying the roof with water. I didn't realize he was joking until he started laughing."

Adrien doesn't get rattled. But what followed early that Tuesday afternoon shook her to the core. She was at home with Kamille, the four-year-old Congolese boy she and Lucas had recently adopted. Justice, her six-year-old, was at school. After seeing pictures of the fire on Facebook and the television news, Lucas's sister called from out of town, urging Adrien to leave. Adrien declined, assuming that local officials would tell her if that was necessary. "Kamille had gone through a lot before we finally brought him to Fort McMurray in March," she explained. "The adoption procedure had not gone well. His English was not good. He was having a tough time adjusting. I didn't want to worry him all over again by moving away. Home, I thought, is forever. And this is where he needed to be."

Like many others in town, Adrien still didn't appreciate the perilous nature of the situation until the school called, telling her to pick up Justice. On the way there, she saw people gassing up cars packed full of personal belongings. "'These people are crazy,' I thought. I was convinced this was not going to be a big deal. I didn't know that some parts of town were under an evacuation alert." At the school, teachers were openly talking about the city burning down. Adrien still didn't think it was as big deal as that, but decided to go home and start packing just in case.

"As soon as I started to put things into bins, Kamille had a meltdown and started to cry. I didn't know what to do except to put my mom on the computer so she could talk to him and calm him while I packed. I then called Lucas just as he was trying to call me and told him that it looks like we will have to evacuate. 'Yes, you should – right now,' he told me." Adrien assumed they would only be gone for a day, but decided to pack for three days just in case. Along with clothes and toys, she brought plenty of food and snacks, including peanut butter and packages of Kraft Dinner, the kids' comfort food. In the car, she looked around and figured that things didn't look so bad. Once again, she wondered whether she might be overreacting.

20.3 RCMP officers were dispatched to direct traffic out of Fort McMurray.

"I called our friend Matt Miniely, who is a pastor at our church. He told me not to even consider staying. He insisted that we meet him at his sister-in-law's place in Timberlea, which is close by, so that we could all leave town together." Just as she was about to pull out of the driveway, a neighbour advised Adrien to go right instead of left on the street because she was likely to run into a firefighting crew blocking the way. Had she not done what he recommended, she and the kids would have seen Lucas and his crew trying to save homes burning in their neighbourhood. "I think we would have all had a meltdown if that had happened," she said.

By this time, the gridlock that characterized the mass exodus from Fort McMurray was already well under way. Traffic was terrible, and it appeared that no one was going to let her merge from the city into one of the highway lanes, until someone kindly did just that. "And that's when I see this guy sitting in the back of his van on the side of the road playing a guitar," she says. "It was a *Titanic* moment. 'We're all going down,' I thought to myself. 'Everyone is going down.'"

There are a number of reasons why no one "went down." Demographics was one of them. No one goes to Fort McMurray to retire. So, not a lot of time and effort was spent trying to deal with the aged and infirm who might not have had a car or the wherewithal to drive one down a crowded highway through black smoke with embers shooting across their windshields.

The other reason had to do with the nature of the people who lived in Fort McMurray. Most were employed by the oil sands industry, where rigorous

20.4 There were not enough masks, two-way radios, or police cars for all the RCMP officers.

health and safety training programs remind everyone to listen to the authorities in times of emergency, even if you think they are wrong. Employees lose their jobs if they don't comply.

RCMP constable Andrew Brock was thankful for this co-operative attitude when he was dispatched to the four-lane highway intersection in the neighbourhoods of Beacon Hill and Gregoire to direct traffic. The toughest part, he said, was stopping worried parents from going to pick up their kids at school. It was the big burly fathers in pick-up trucks who really worried him. "Imagine what you would say if I told you that you couldn't go into a burning neighbourhood to get your kids?" he asked me as we stood at the intersection to review what had happened that Tuesday afternoon, the day of the evacuation on 3 May 2016. "You could see from where we're standing how the smoke and flames were burning trees and buildings around the school up there. I'm telling the parents that they can't go in because I need to get everyone out and to let the fire trucks in. I ask them to trust me that their kids are being loaded onto a school bus and being driven to Anzac south of town. The thought occurred to me more than once that I would be run over. I'm not sure I could have blamed them if they did," he said.

RCMP efforts were further hampered by the fact that they did not have enough cars and radios to go around. Those radios they did have had no direct connection to the firefighting teams. Masks ordered at the last minute didn't arrive in time because the truck that was bringing them in broke down along the way. And then there was the issue of the morgue that Dicks was certain was going

20.5 The exodus of cars and trucks heading into smoke during the evacuation of Fort McMurray in 2016.

to be needed. She had no idea what to with hundreds, potentially thousands, of bodies that might be discovered the next day.

But residents of Fort McMurray also got lucky because the twinning of the highway going south had just been completed, allowing cars to get out more quickly than they would have with a single lane. There had also been a lull in the oil sands industry that resulted in several camps being empty, but with enough food and water to feed and house those residents who were sent north to oilsands facilities to prevent further delay in getting people out of town as quickly as possible. There is only one road that passes through Fort McMurray. The road going south was bumper-to-bumper. There was not much traffic going north to the oilsands.

Finally, the weather cooperated when shifting winds drew the fires away from town just as water mains stopped pumping and as firefighters were engaged in a form of triage, knocking down one burning house to save another next door to it. Chaos and confusion prevailed until the emergency response team finally ordered the evacuation that was already well underway. None of the reports stated it, but the decision should never have been left to a regional emergency response team, many of whose members were preoccupied by what would happen if the oilsands didn't have enough employees to keep things running. A shut-down and a start-up would eat into a year of revenue. It might have been ugly had 88,000 people fled a fire that did not come. The decision should have been made at the provincial level, where a task force involving every government department had been in play since the day the fire started making a run towards the city.

Jamie Coutts, the Slave Lake fire chief, didn't pull any punches when he met with government officials who called a post-mortem on what went wrong. He echoed virtually everything that the review panels had noted but salted it with some pretty tough words. If there was a lesson to learn from this costly disaster, he said, it's that there needs to be a unified chain of command when urban and forest firefighters are compelled to work together, and that evacuations in perilous situations such as this one need to be ordered earlier, even if there is pushback from the politicians, the public, and industry.

The son of a firefighter, Coutts began his firefighting training when he was just fourteen years old. As resilient as he was to the tragedy that often comes with fire, he has never been able to shake off the sight of seeing so many houses owned by people he knew burn down to the ground in Slave Lake.

"The thing that struck me most when I was in Fort McMurray while the fire was burning was how similar this was to what happened in Slave Lake. Seeing all that chaos was really demoralizing. It was like Slave Lake never happened. We didn't learn anything. I came away thinking that bad as it was, Fort McMurray dodged a bullet. This is going to happen again."

20.6 The total cost of the 2016 Horse River Fire was close to $9 billion.

21

NUCLEAR WINTER: PART TWO

Four days into the Horse River Fire, a massive, anvil-shaped tower of smoke rose eight kilometres into the powder-blue sky, hurling thunderbolts far from the fire's flanks. Those who were tracking the progress of the fire with satellite imagery were startled when they saw how that lightning ignited a cluster of blazes thirty-five kilometres away.

Cordy Tymstra, the fire science coordinator for the Alberta government, had just written a book on the massive Chinchaga Fire that ripped through this region in 1950.[1] He thought he understood the power of such conflagrations. But this angry dragon-like behaviour was something rare and unexpected on a day when there had been nothing but blue skies. "It was a lightning storm generated by that fire cloud that was responsible. This does happen," he told me when we met months later to talk about what had happened that day. "But I have never heard of lightning causing new fires so far in advance of the main fire."

The physics behind pyrocumulonimbus clouds, or "pyroCbs," are complex and continue to puzzle meteorologists. What we know is that the heat of an intense fire creates an updraft which sucks smoke, ash, burning materials, and water vapour from lakes, streams, burning vegetation, and the atmosphere high into the sky where they cool and form fire clouds that look and act like those associated with classic thunderstorms. One by-product of the combustion process is the water vapour that adds energy to the development of pyroCbs.

In extremely rare cases, these fire clouds can produce rain. Almost always, though, the heat and the particulates in the smoke trigger a chemical reaction

21.1 As the climate warms, lightning strikes are increasing and extending their reach into the far northern regions.

that arrests the ability of the cloud to produce precipitation. What's left then is a lightning storm that stokes the surrounding landscape, triggering more fires, as it did near Fort McMurray in 2016.

In any case, a form of meteorological chaos ensues. The cooler air that rushes in to replace the rising hot air causes the fire to shift in unexpected directions, often with powerful gusts that can roll over incoming winds or speed up those that are pushing the fire forward. Updrafts can also collapse, sending hot ember-filled air downward where it hits the ground and spreads in all directions.

Scientists had for some time speculated about the possibility of an intense fire resulting in Pyro-tornadogenesis, a term used to describe the ability of a large fire to create a tornado. Most were either sceptical or sitting on the fence until 18 January 2003, when a tornado formed in the plume of the McIntyre's Hut Fire that crackled through the Canberra region of southwestern Australia. Even then, many remained doubtful, believing that what was seen was not a tornado but a less powerful phenomenon known as a fire whirl.

Rick McRae, the lead scientist in a study into that fire's behaviour, ended the speculation when he and his colleagues confirmed that a tornado had formed. They gave it a rating of at least a 2 on the Enhanced Fujita scale of tornado severity. "It had major effects on the behaviour of the fire on the urban edge and had enough force to remove roofs from houses and to blow cars off the road," McRae told journalists when the study was published nine years later, after much investigation. "It moved at over 30 km/h across the ground and had a basal diameter of nearly half a kilometre when it reached Chapman. It was a major tornado, but was barely noticed given the setting," he said.[2]

The Canberra fire tornado, the Fort McMurray pyroCb, and several other extreme fire events were considered to be so anomalous that they were more talking points than causes of genuine concern for firefighters. But events like these have staged an astonishing run since 2013, when the Rim Fire in California produced at least two large pyrocumulonimbus (pyroCb) events in what was one of the most intense fires in California history.[3] The first of a number of those head-spinning pyroCbs took place on the evening of 12 August 2017, when heat and smoke from an intense wildfire burning in the forests of British Columbia began mushrooming skyward, sucking up ash, blazing wood and vegetation, and water vapour from lakes and streams below. Rick McRae was in British Columbia that evening helping Canadian colleagues manage the fires that eventually drove more than 65,000 people from their homes. Sensing that the conflagration was going to erupt into something extraordinary, he texted a group of scientists from around the world who had been collaboratively studying fire-triggered thunderstorms since the 2013 Rim Fire in California. "This is a very bad day all around in western Canada," McRae wrote. "Fires went 'pop' progressively on the leading edge." McCrae went on to say that a pyroCb was forming in the southeast corner of British Columbia, near Kamloops, adding "Things could get worse if certain things 'align.'"

And align they did. As fires that would eventually consume 1,217,294 hectares in the province burned out of control, four fire-driven thunderstorms rose over the conflagration, shooting black smoke and carbon high into the lower stratosphere, spewing noxious gases that were eventually detected almost as far north as Greenland and Ellesmere Island in the High Arctic and touching off more fires in BC. At the same time, fires in neighbouring Washington State spawned yet another pyroCb. "By later that night we were agape about the cluster of pyroCbs and the extremely impressive smoky anvil-shaped clouds," Mike Fromm, a meteorologist at the US Naval Research Laboratory, told me shortly after the study he co-wrote was released.

Fromm and David Peterson, a colleague at the Naval Research Laboratory, were so impressed they labelled that day's cluster of pyrocumulonimbus cloud activity "the mother of all pyroCbs," surpassing even the events of Black

21.2 Pyrocumulus cloud, the Beaver Fires Complex, 2014.

Saturday in Australia in 2009, when monstrous bushfires killed 173 people, injured 414 others, and burned 4,400 square kilometres of bush. On that night in February, three clearly distinct pyrocumulonimbus storms erupted across the state of Victoria in southeastern Australia. On 7 February, the largest of the plumes reached a height of fifteen kilometres, with enough energy below to generate hundreds of lightning strikes.[4]

I was in Washington, DC, in December 2018, both as speaker and listener, when a packed room of scientists discussed the significance of these storms at the annual American Geophysical Union meeting. They considered the possibility that firestorms like these could act like volcanos and maybe cool the world as Mt Pinatubo did when it erupted in the Philippines in 1991. Some researchers even recalled the "nuclear winter" theory that Richard Turco, Carl Sagan, and others had posited more than thirty-five years earlier. No one was predicting anything quite as dark and destructive as that, but they were considering the possibility that pyroCbs could damage the ozone and maybe result in some regional short-term cooling.

What was needed, of course, were more pyroCbs to study. No one expected they would arrive so quickly and with increasing frequency, as they did in the months and years that followed. Scientists were seeing fire-driven thunderstorms

21.3 Pyrocumulus cloud.

in places such as Texas, Portugal, South Africa, and Argentina, where they had never occurred before. Destructive as the Canberra fire tornado was in 2003, the Carr Fire in California in 2018 produced one that was even more powerful. It came with wind speeds of up to sixty-four metres per second, equivalent to an EF-3 tornado.[5] That's the kind of tornado that can overturn a train, rip the bark off a tree, and severely damage a well-constructed shopping mall. But, to the continuing astonishment of everyone involved in tracking pyroCbs, there was much more to come.

The biggest and most frightful pyroCb show occurred in the winter of 2019–20 when Australia suffered through its worst fire season ever, which is saying a lot given the toll that the country's bushfires have taken since the 1974–75 fire season burned 117 million hectares – equivalent to the area of France, Spain, and Portugal combined.

The fire was all consuming, burning more than eighteen million hectares and killing more than three billion animals, according to some estimates. It would have been unimaginable had California not burned as big and intensely as it did in 2017 and 2018. What made the Black Summer Fire really stand out were the eighteen pyroCbs that erupted between 29 December and 4 January. It was a meteorological show like nothing ever seen before – the kind you expect in over-the-top sci-fi films like *The Day After Tomorrow*, which depicted the rapid global cooling that takes place following the disruption of warm water circulation in the North Atlantic Ocean.

The largest of the plumes in Australia, according to Fromm, David Peterson, George Kablick, and four other scientists exhibited several previously undocumented phenomena.[6] No other fire in history had sent so much smoke to heights of up to sixteen kilometres. No other plume had exhibited such remarkable lofting. One plume rose from fifteen kilometres to thirty kilometres. It was five kilometres thick and 1,000 kilometres wide. It travelled east from Australia to South America before reversing course in January and completely circling the globe westward over the next few weeks. Some of the smaller plumes were rising and rotating wildly. Remarkably, the plumes shaded out the rays of the sun long enough to cool the global climate by about .06 degrees Celsius. According to the study that documented this cooling, it wasn't so much the ash that caused it, it was the physical modification of the cloud cover that came with it.[7]

According to Fromm, Kablick, and their colleagues, this was the first evidence of smoke causing winds in the stratosphere to change direction. It opened a whole new vein of scientific research. To suggest that almost no one saw any of this coming twenty years ago speaks to the fact that history of fire is no longer a proxy for what is likely to happen in the future. Until the turn of the century, most scientists didn't think fires had enough energy to do any of this, and many scoffed at those who floated the possibility.

Mike Fromm saw a bit of this pushback when he began working for the US Naval Research Lab (NRL) in Washington, DC, in 1993. The NRL provides the advanced scientific capabilities required to bolster the country's global naval leadership. In pursuit of this mandate, some of its scientists explore the physics behind hurricanes, volcanoes, and wildfire. The Naval Research Lab had sent a satellite into space the year Fromm began working there. The Polar Ozone and Aerosol Measurement data was monitoring the stratospheric clouds and plumes that were the subject of the first papers he wrote.

Fromm's interest in pyroCbs began in the summer of 1998 when the satellite instruments he was using homed in on weird-looking clouds in the stratosphere. Weird, because clouds like those were thought only to form in winter. However, this wasn't the first time that scientists had seen such clouds in the summer stratosphere. Puzzling images of aerosol clouds had appeared periodically over the boreal regions in the summers from 1989 to 1991. But the scientists who identified them then concluded that they must have come from volcanic eruptions because they, like everyone else in the scientific community, believed that thunderclouds formed by wildfire didn't have the energy to get past the troposphere, which contains half of the Earth's atmosphere and where weather occurs. There was just one problem, however, according to Fromm. There was no documentation of a volcano erupting anywhere in the world during that time.

Try as hard as he could, Fromm found no evidence of volcanic activity to account for what he was seeing in the stratosphere. Nor could he convince his colleagues that something was amiss. And that's when he began doing some detective work to determine whether pyroCbs might be responsible. Among the first people he called was Canadian meteorologist René Servranckx of the Canadian Meteorological Centre in Dorval, not far from Montreal. Servranckx, according to Fromm, was crucial to making the case for the pyroCb explanation. "Many scientists who saw our early presentations scoffed," he recalled when we spoke at the AGU conference in Washington, DC. "One actually chuckled at the idea. That might have been a seminal trigger to get to the bottom of this wild weather story."

Fromm eventually presented his theory to Brian Stocks of the Canadian Forest Service. Stocks had been conducting research on fire behaviour in the boreal since the 1970s. He had shown atmospheric scientists like Richard Turco how intense fires behave, and how high their plumes rise. When Stocks asked Fromm to remind him at what height the stratosphere began, Fromm told him that it was between 25,000 and 30,000 feet. Stocks told him that he had measured many columns of smoke that had travelled higher.

That was the start of some great collaboration between Fromm, Stocks, Servranckx, and many other wildfire and atmospheric scientists. Using satellite image data, satellite-based profile data, and ground-based lidar data (a technology used to create high-resolution models of ground elevation), Fromm eventually linked those weird clouds he detected in the stratosphere during the summer of 2001 to extreme pulses of smoke from the Chisholm fire in Alberta, the one that set a new record for intensity on Canada. The smoke from that fire drifted around the world more than once.

The most powerful insights occurred in 2008 when Fromm, Stocks, and more than a hundred scientists from around the world descended on a NASA-led research base site at Cold Lake, Alberta to track smoke from wildfires burning in western Canada. There had been more than 460 fires early that summer. One of them was an enormous, fast-moving blaze that had the RCMP scrambling to evacuate more than 2,000 people out of northern Saskatchewan communities like Uranium City, Black Lake, Stony Rapids, and Stony Lake on the third day of July.

On the fourth of July, Fromm and a small group of scientists saw their opportunity and made the first-ever flight into an active pyrocumulonimbus cloud to get measurements in the updraft core. There were no fireworks to welcome the Americans. But unlike conditions found in the core of any other cloud, this one produced total darkness, thanks to the vast abundance of smoke and small cloud droplets. The possibility that wildfire could send smoke from the troposphere where the pollution it generates is a regional problem into the stratosphere where it becomes a global problem was underlined on that day. Flying into what they suspected was a pyroCb, Fromm could smell forest fire smoke at 34,000 feet inside the NASA aircraft.

The more Fromm and his colleagues looked, the more links they discovered: the 1988 fires in Yellowstone, the 1989 fires in Manitoba and Saskatchewan, the 2003 Conibear Fire in the Wood Buffalo National Park in the Northwest Territories of Canada – all of these had produced pyroCbs with enough energy to pass through the troposphere into the stratosphere. One of the most impressive was the 1998 Norman Wells Fire that produced four enormous pulses of smoke just south of the Arctic Circle. Another was the 2002 Hayman Fire in Colorado that was started by a US Forest Service employee who was allegedly burning a letter from her estranged husband.

"PyroCbs from the Circle Fire in Alaska explained the stratospheric anomalies that were reported in 1990," Fromm told me. "We also now know that the Yarnell Hill Fire in Arizona in June of 2013 spawned a pyroCb that was active at the time nineteen men were killed fighting that fire."

As recently as 2017, Fromm doubted that pyroCbs could influence climate in the dramatic ways that major volcanic eruptions like Mt Pinatubo have done. Aerosol pollutants from that eruption in the Philippines reduced global temperatures by one degree Fahrenheit over a period of fifteen months. The sea ice in western Hudson Bay lingered for so long the following spring that polar bears got extraordinarily fat because of the few extra weeks they had to hunt seals. Seals represent 90 per cent of a polar bear's diet and the most efficient way of killing them is on the sea ice. In a research paper explaining the cause for the weight gain, Ian Stirling, a Canadian Wildlife Service scientist, dubbed them the "Pinatubo bears."

But given all that has happened since that flight into a pyrobCbs in 2008, Fromm was not so sure that pyroCbs weren't as impactful as volcanic eruptions. The thing to watch out for in the future, said Fromm, is what will happen if we see more of the multiple pyrobCbs events like the ones that occurred in 2017 in British Columbia and Washington and the ones that occurred in Australia in 2019–20. "One pyroCb may not have the weather- and climate-altering capacity of a significant volcanic eruption," he told me. "But several clusters occurring in a year might," as they ultimately did during the 2019/2020 Australian bushfire season.

22

OWLS AND CLEAR-CUTS

On a hot May day in 2019, fire ecologist Chad Hanson and I were hiking through the Stanislaus National Forest, heading toward an empty great gray owl nest he found earlier that spring. This type of owl is genetically distinct from its cousins in western North America. It is two feet tall, with a wingspan of about five feet. It can be seen almost any time of day because, unlike most other owls, it is active day and night. There were at the time only 200 to 300 adults remaining in California where forests have been battered by drought, insects, rising temperatures, and wildfires like the Rim Fire that burned in 2013. In the three years that Hanson – principal ecologist with the John Muir Project, a non-profit group devoted to the ecological stewardship of federal forests – has been conducting research in the Sierra Nevada, he had yet to see an adult gray owl.

The Rim Fire got its start on 13 August 2013, when deer hunter Keith Emerald was in the Clavey River watershed of the Stanislaus. Officials had posted a forest-wide fire ban weeks earlier. But like Thoreau, when he and a friend were fishing in Massachusetts in 1845, Emerald was hungry for the soup that he had brought along. When he lit a fire to heat it up, gusting winds transported embers uphill into the tinder-dry trees that had been baking under extraordinarily intense temperatures all summer long. Before Emerald realized what had happened, the fire was out of his control.

The Rim Fire raged in the Stanislaus and along the boundary of Yosemite National Park for nine weeks, shutting down park roads and prompting thousands of visitors to cancel vacations. More than 5,000 firefighters were

22.1 Post-fire salvaging in the Apache National Forest. Photo by Lance Cheung.

brought in at a cost of more than $125 million. One hundred thousand hectares of forest was charred. It was California's third largest and most costly wildfire up until that time, and yet another of those conflagrations that suggested that the future of fire would not necessarily be shaped by the strict laws of nature but by social and political choices, reckless human behaviour, and never-ending increases in greenhouse gas emissions. There could be no denying the relationship between high temperatures, low humidity, and large fires. The Rim Fire was one of California's largest between 2000 and 2021. Eleven of the twelve warmest years on record occurred during that time.[1] The heat that helped ignite them was largely generated by humans.

As federal agencies in the US cut costs to deal with a succession of budget standoffs such as the one that occurred as the Rim Fire burned in 2013, there was further pressure to decrease the size of a firefighting workforce that had already been reduced by 40 per cent since the 1980s. Increasingly, the cost of fire suppression was being forced upon state and local jurisdictions. California, Arizona, and Nevada responded by hiring thousands of prisoners who were paid between fifty cents and a dollar per hour to fight fires. And increasingly, the states and the Forest Service opened the door to salvage logging – the practice of logging trees in forest areas that have been damaged by wildfire, flood, severe wind, disease, and insect infestation – to produce the revenue needed to revitalize the forest, reduce the risk of future fires, and ensure public safety. In 2016, the Forest Service sold $186 million of Rim Fire timber.

This kind of salvaging was and continues to be an experiment that has either failed miserably or succeeded nicely depending on who one talks to on both sides of the US-Canada border. It's a century-old debate that goes back to the era when Gifford Pinchot led the US Forest Service and when Elihu Stewart headed up what would eventually become the Canadian Forest Service.

As Hanson and I moved through the forest into a meadow that day, we spotted a great gray owl in a snag – a bare, fire-scarred tree. Hanson was elated, not so much for the opportunity to finally see an adult of this species in the Stanislaus, but because it drove home a point he had been making about how this fire, and many intense fires like it, did not "nuke" the landscape as a US Forest Service scientist suggested,[2] and as the federal agency stated in softer language in a series of online videos. "Since the Rim fire burned, we've had this steady stream of people insisting that this was one of the most catastrophic and devastating fires in California history," Hanson told me. "They claimed that nothing would grow back, that the owl and deer populations could not be sustained. They, and others, are using it as an excuse to accelerate the clearcutting of snag forests."

In a climate change era in which wildfires are beginning to burn bigger and more intensely than they have over the past century in places such as California, Colorado, British Columbia and other places, there is genuine concern that the continent's forests will not regenerate as they did in places such as Yellowstone, which many people wrongly assumed would be destroyed by a 1988 conflagration that famously ripped through the park. Parsing out right from wrong is challenging, because the scientific literature on post-salvage logging is rife with contradictions, as US Forest Service scientist Joseph Wagenbrenner concluded when he was at Michigan Technological University until 2015.[3] Some say it is beneficial because it churns up the ground, removing water-repelling soils that sometimes form after an intense fire. Proponents also insist the detritus left behind inhibits erosion. But critics such as Hanson argue that the skidders – heavy vehicles used to transport felled trees to the trucks that carry them out – kill seedlings and compact the soil in a way that increases runoff and erosion and decreases natural regeneration. Wagenbrenner's own research shows that salvage logging can lead to potentially lethal increases in sedimentation and water temperatures in streams.

The Forest Service did not get all the revenue that it wanted from the Rim Fire because the Center for Biological Diversity and the John Muir Project, an organization that Hanson co-founded, went to court in August 2014 to stop the clearcutting. They lost the case. But the delay was long enough to dissuade

22.2 Elk and moose do well in post-fire ecosystems feeding on the shoots of young aspen. Photo by Edward Struzik.

the US Forest Service from completing the prescribed clear-cuts because the value of the timber had decreased with decay.

The snag forest the Rim Fire produced, according to Hanson, is good for the ecosystem. Great greys such as the one we saw are nesting on the edge of this high-intensity burn, hunting for rodents that are attracted to open meadows and burned-out areas like the ones we hiked through. Black-backed woodpeckers, a species that is being considered for listing under the Endangered Species Act, are thriving on fire beetles that lay their eggs on snags. These larvae get fat feeding on the decaying wood. Other creatures such as mountain bluebirds, western screech owls, flying squirrels, and fishers exploit the woodpecker-drilled cavities for their own nests. There are more spotted owls in the Stanislaus Forest now than there were before the fire.

"Snag forests are as ecologically important as old growth forests and other forest ecosystems," Hanson told me that day in the Stanislaus Forest. "But there is no protection for them." There is ample evidence that most birds and animals benefit from the after-effects of fire, as grizzly bears, elk, moose, and other species do in the Rockies and throughout the boreal forest.

But as fires burn bigger and more intensely in places such as California and Australia where there are now shorter intervals between high-severity fires, regeneration is being compromised. Hotter fires that burn more frequently make it very difficult for fire-tolerant trees such as eucalypts and alpine ash to regenerate. These trees – which need to be fire-free for twenty years in order to produce seeds for successful regeneration – haven't been getting that growing time in eastern Victoria. Since 2003 there have been, in the space of just seventeen years, four intense fires. The same thing is happening in parts of the boreal forest of Canada and Alaska where conifers such as spruce and pine are no longer as blessed with long fire-free intervals, as they once were, and are simultaneously being cursed by higher temperatures, rapidly thawing permafrost, drought, and disease.

Intense fires that burned Wood Buffalo National Park in 2004 and again in 2014 may be a harbinger of what's to come. Since the second fire, there is little evidence of aspen and pine popping up as they did after the first. In the unlikely event that strong winds blow in pine seeds from far-off unburned areas, they are not likely to germinate because the fires' heat vaporized most of the organic matter in the soil. "That was a pine forest for centuries," Marc-Andre Parisien, a research scientist with the Canadian Forest Service, told me in an interview in his office in Edmonton. "It may be decades or even centuries before it's a pine forest again."

Ellen Whitman worked on the Wood Buffalo site under the supervision of Parisien and Mike Flannigan when she was a PhD candidate. She says the outlook in Wood Buffalo looks a lot bleaker than it might otherwise be because of a period of intense drought that followed the second fire. Whether the pine will return any time soon, she told me in an interview, is uncertain but doubtful.

Two things happened that have made pine forest regeneration unlikely. To have a truly successful regeneration, you need to have a good seed crop. For that to happen, the pine trees have to be ten to twenty years old, depending on stand characteristics; some say they have to be much older. In this case, the trees were just ten years old when they burned.

Even if there was a good seed crop, the seedings would need some sheltering sites and water, and maybe shade from time to time, to grow. The fire in 2014 was so severe that it came close to completely removing what was left of the organic soil layer. Seeds are trying to grow on sand that doesn't hold much moisture. Some vegetation is coming back, but there is almost no sign of trees. They wouldn't have been able to produce a seed crop that is necessary for the forest to grow back the way it was before the first fire in 2004. Hanson and I saw a bit of this in the Stanislaus, where the US Forest Service planted seedlings in the clearcut. Many were dead. Most showed signs of extremely slow growth. They

22.3 Some fires burn so hot and so deeply into the duff that there are not enough seeds or nutrients remaining for trees to regenerate. Photo by Ellen Whitman.

didn't have the benefit of shade nor the moisture they needed because most of California was at the tail end of a long catastrophic drought.

This combination of hotter summer weather, drought, and more frequent fires may even bring an abrupt end to the Ponderosa pine and Douglas fir forests that dominate the lower elevations of western North America. Kimberly Davis and her colleagues came to this conclusion when they looked at what followed thirty-three wildfires that burned in several regions in the western United States. It was similar to what Whitman saw in Wood Buffalo. The intensity of the fires combined with drought, and the lack of seed availability reduced the opportunity for seedlings to establish.[4]

Hanson conceded that not all forests are equal in the way they respond to fire. But he doesn't see this kind of succession happening in the Stanislaus and most other burned forests in other parts of the American West.

The deeper we travelled into the burn, the more verdant and biologically diverse the vegetation became. We saw Douglas Fir, Ponderosa pine, sugar pine, and black oak rising up from a forest floor carpeted with white leaf manzanita and mountain misery, a highly aromatic shrub that oozes sticky black gum. A red-tailed hawk gave us as a start as it swooped by in an open meadow. Hanson pointed to one cluster of five-foot-tall pines, veritable giants compared to the seedlings that the Forest Service planted in its clearcuts after spraying herbicides to reduce competition for water, nutrients, and sunlight. "Basically, clearcutting and spraying herbicides amounts to kicking a forest when it's down and trying to regenerate on its own," Hanson said. "The heavy machinery used to cut down the snags destroys the conifers that are trying to grow naturally. The herbicides prevent the growth of native shrubs and forbs (herbaceous flowering plants that do not include grasses) that are beneficial to animals such as deer and elk and bears that feed on berries. They favour instead invasive species such as cheatgrass, which is stress tolerant."

Hanson didn't know exactly how and why the vegetation was returning so nicely in the intense burns. He believed the Forest Service was sincere in concluding that nothing would grow back. "Squirrels and birds could have brought the seeds in from other areas. A rapid snow melt on steep slopes in the springtime could have transported the seeds in with a flood. The fact is we don't really know. But it's pretty obvious that the Forest Service was wrong."

Hanson stopped to show me a snag pock-marked with hundreds of holes pecked by black-backed woodpeckers. "One black-backed woodpecker needs to eat 13,500 wood-boring beetles a year in order to survive. A pair of woodpeckers supporting chicks needs 100–120 hectares of snag forest. Without these snags, we would not have the diversity of insects we have, nor the woodpeckers and other wildlife that depend on them."

Richard Halsey, the director of the California Chaparral Institute, was a member of Governor Jerry Brown's Fires Services Task Force on Climate Change that was established May 2018. He was discouraged by the decision to accelerate salvage logging because there is so much scientific evidence weighing against it. Treating nature as fuel, he suggested when I interviewed him shortly after my hike into the Stanislaus, is dangerous because it disregards the fact that forests protect water supplies, temper climate warming, and promote biodiversity. "My conclusion is that this is all about money," he said. "This focus on dead trees is misguided because the fires that burned in California had nothing to do with forests. I'm all for removing hazard trees near homes and roads. But it makes no sense to spend millions of dollars cutting down trees and snags when the real problem is people starting fires and people building highly flammable homes in forest and shrubland environments."

23

WATER ON FIRE

Cameron Falls in Canada's Waterton Lakes National Park runs cold and clear in summer when as many as a half million people come to canoe, fish, hike, and bike in this pristine mountain landscape along the Alberta-Montana border. On very rare occasions, it runs a Pepto-Bismol pink when heavy rains stir up argillite, a red mudstone that is found upstream. But on 21 June in 2018, residents, tourists, and park officials were surprised to see the waterfall suddenly running pitch black. Heavy rain had flushed in soot, ash, and charred tree debris from a fire that burned most of the park the year before.

When it happened again a few weeks later following another violent thunderstorm, the dark flow of mountain water rushed in with so much ferocity that it piled charred branches and debris into the gorge, into a neighbouring parking lot, and onto a street in the townsite. It was then that people realized that serious water quality issues might be coming for the park townsite, for fish and other aquatic animals, and for ranchers and residents living downstream.

Before the Kenow Fire, there were trees in Waterton that were already alive when two members of David Thompson's fur-trading party passed through in 1800 with Kootenai Indigenous guides leading them. Fire was not as common a visitor here as it was in Glacier National Park to the south, the heavily forested Kootenay River Valley to the west, and much of western North America. Three relatively small fire events, the Cameron Creek and Oil City fires of 1919 and the Boundary Fire of 1935, accounted for 60 per cent of the area burned between

1910 and 1998 when the last fire of significance consumed just 600 hectares. By way of comparison, 55,000 hectares burned in Glacier in the summer of 2003.

Waterton came very close to burning big in 2015 when a fire in Glacier began heading north, compelling park superintendent Ifan Thomas to order an evacuation alert for the town site. Thomas was taking no chances back then. He had been in Jasper in 2003 when the Syncline Ridge Fire forced an eight-hour highway closure on the eve of a folk festival and what promised to be the busiest long weekend in the park. He was one of a small number of people who knew just how close that fire came to making a twenty-kilometre run to the town of Hinton or back towards Jasper, if the winds had shifted.

Scott Murphy was the wildfire management officer in Waterton in 2017. He knew a big fire was a distinct possibility that summer. Only half the rain that typically drenches the park in June had fallen. A fire ban was put in place on 15 July and stayed as the summer drought persisted throughout August into September. It had been this dry only twice in the park's 122-year-old history.

The long, nervous wait for a big fire ended on 8 September 2017, when a lightning strike in the Akamina area of southeastern British Columbia moved into the park from the west. This could have ended very badly had lessons not been learned from Fort McMurray the year before and from the Reynolds Fire in Glacier that threatened to move into the park in 2015. Unlike emergency response officials in Fort McMurray, who had waited for the fire to come to town before issuing a general evacuation order, Parks Canada officials got everyone out of the park eighteen hours before the fire began its spectacular run. Townsite residents and cottage owners might have balked had it not been for Murphy and his colleagues laying it all out for them well in advance at a community meeting.

No opportunities were wasted to deal with this conflagration. The day after, personnel ready to help arrived from across Canada, including helicopter pilots and structural fire crews from regional municipalities such as Calgary. All but one of the park's bison that were being held in a paddock were shipped off to Grasslands National Park. And yet there was nothing anyone could do to stop this fire from doing what it wanted. Waterton is not the most ideal place to manage a fire. It is mountainous and there are not many wetlands to slow a fire's progress. A fire burns uphill faster than it does on flat land because the flames can more easily reach fuel in front of the fire. Winds powerful enough to make the historic Prince of Wales Hotel shake can drive fires like this one forward even faster.

And that's exactly what happened. During its most intense phase, the fire danced and swirled twenty-six kilometres from one end of the park to the other in just five hours. "It made its move towards the townsite at nighttime, when fires typically calm down," Murphy told me when I was in the region that winter to give a talk on *Firestorm*, my book on wildfire. "I couldn't believe it when I saw

23.1 Soot from a fire in Waterton National Park in 2017 blackened the waters of the Cameron River and Cameron Falls.

that big wall of flames come down the valley and wrap around two mountains. I felt sick realizing we had a crew on the highway that was potentially in harm's way trying to make a final stand. Because it was nighttime, we couldn't fly and get a big picture view as we normally would during the day."

No one was hurt, and most of the homes and businesses were saved. But the fire was so hot and so all-consuming that it ended up incinerating half the vegetated area in the park, three quarters of which burned with extreme severity. The denuded mountainsides and copious stands of dead trees created so many hazards for backcountry hikers that park officials closed 80 per cent of the trails the next year when Cameron Falls was running black for a day.

Uldis Silins, a University of Alberta scientist who has studied several watersheds intensely burned by fire, told me that he'd never seen such blackness

when he was called in to survey the situation. It was August 2018, a little over a year after the fire burned. We were driving into the park with the intention of hiking into the closed-off areas to check on some water quality and water flow instruments he had installed along the Cameron Creek, which flows from a lake with the same name on the Montana border to the Waterton townsite. Silins was confident that the spruce, pine, and aspen would recover as they typically do after a wildfire. What concerned him was the possibility that this fire would have a deleterious impact on fish, on drinking water, and on the denuded mountains slopes that are vulnerable to the kind of erosion that results in landslides and avalanches.

23.2 University of Alberta forest hydrologist Uldis Silins monitoring water quality following the Kenow Fire in Waterton, 2017. Photo by Edward Struzik.

I had already been to Silins's research site in the nearby Castle Crown Wilderness, where an intense wildfire in the Crowsnest Pass in 2003 continued to have an impact on the Lost Creek mountain watershed. He and Monica Emelko, a University of Waterloo environmental engineer, had been monitoring how the ecosystem had been recovering since that fire forced nearly 2,000 people to evacuate the region. But getting a glimpse of this one in Waterton added a more immediate dimension to the role that fire plays in aquatic ecosystems. "See these black mounds of mud on this side channel of the creek [Cameron]. They're rich in nitrogen and phosphorous, nutrients that are driving the growth of this green algae we're seeing over here." As Silins talked, Chris Williams, his associate, scooped up a handful of the bright green, goopy vegetation. "You don't normally see algae growing in these nutrient-poor mountains streams, especially at this time of year, when it is often cold enough to snow in late summer," said Silins. "But that's what an intense fire often does. Ranchers, residents, and park officials were concerned about what this will do to their drinking water and to the aquatic environment. Hopefully we can answer their questions by monitoring water quality in the burned watersheds."

The relationship between wildfire and watersheds is a complex one, each affecting the other in ways that were largely overlooked until the mid-1990s when the first of more than thirty large and intense fires began burning the mountain forests along Colorado's watersheds. The South Canyon Fire of 1994 was most notable for taking the lives of fourteen firefighters on Storm Mountain. It was a relatively small fire, but severe enough in one locale to send twenty-five tons of charred debris down the mountainside onto the Interstate when heavy rains followed.[1]

The 2002 Hayman Fire, which was powerful enough to create a pyroCb, demonstrated for perhaps the first time how an intense fire can affect a watershed. The fire burned through 56,000 hectares over more than three weeks, removing as much as 3,500 hectares at the Cheesman Reservoir, where one of the world's tallest dams was completed in 1905. The soils in that and other denuded areas upstream baked in the dry, hot, conditions that followed. Some spring-fed streams stopped flowing. Chemical compounds vaporized by the fire got driven into the soil. As they condensed, they formed an impervious layer just below the surface. "Hydrophobic" – meaning averse to water – is the word geologists use to describe such soils.

Without trees, vegetation, and a stable soil structure to absorb the heavy rains that followed, tonnes of ash, debris, heavy metals, and nutrients were flushed through the watershed, just as in Waterton. The magnitude of these bursts, however, made the Cameron Falls event look like a squirt from a water gun.

Washed-out highways were the least of the state's problems. The after-effects of the fire led to the precipitous decline of the blue-ribbon South Platte River trout fishery, which was the best of any river in the United States within a one-hour drive from a major city.[2] Worst of all was that the affected watershed provides drinking water to 75 per cent of the state's 4.3 million residents. Hundreds of tons of sediment filled lakes and reservoirs. Intakes got clogged. Water quality suffered not just for a few days, but for fifteen years or more.[3] To find a solution, more than sixty scientists from various disciplines were brought in. Crews dredged out tens of thousands of tons of sediment. Some 175,000 trees were planted. Still, the drinking water quality problems persisted, even as native plants recovered and resisted being taken over by exotic species, as sometimes happens after an intense fire.

Since then, several other US states and Canadian provinces, as well as four major Australian cities, have undergone similar challenges. Denver spent $26 million (US) to restore the function of water storage and collection.[4] Canberra spent $38 million (AUD) after the 2003 fire.[5] The Rim Fire of 2013 came so close to threatening the water supply for 2.6 million Californians in and around the San Francisco region that officials were scrambling to figure out how to divert water from other sources in case the Hetch Hetchy reservoir was overwhelmed with soot and ash. Paradise, California, wasn't so lucky when the 2017 Camp Fire resulted in widespread chemical contamination of the water distribution system. Silins and Emelko were brought in to help Fort McMurray to cope with the huge amount of toxins and sediment that continued to challenge the city's water treatment plant six years later.

One of the biggest ongoing problems was the dissolved organic carbon released by a wildfire. When mixed with the chlorine often used to treat water, it can produce carcinogens that most treatment plant technicians don't have the expertise to manage. To deal with this challenge, Fort McMurray was continuing to spend more than twice as much on treatment as it did before the fire because standard technicians didn't have the expertise to ensure that the water was safe.

With fires burning bigger, hotter, and more often, threats to water supply are bound to escalate, according to Deborah Martin, a Colorado-based US Geological Survey scientist. An increasing number of regional, national, and global water assessments are now including wildfire in evaluating risks to water supplies. "Forests yield 40 per cent of the water for the world's one hundred largest cities," she told me when I interviewed her about the study she had published. "Many of these cities are already water-stressed because of drought, climate change and increasing water consumption. The increase in wildfire activity could make it much worse."[6]

The situation in Canada typifies the challenges that many countries face if fire continues to burn more often and intensely. Two thirds of the country depends on water that is stored and filtered in forests.[7] In many places, the quality of that water is already being degraded by drought, pollution, climate change, agriculture, and urban development. Groundwater may be keeping cool and clean in places such as Lost Creek and possibly in Waterton National Park, but no one in Canada knows for how long because few jurisdictions and agencies have adequately mapped out, evaluated, and diligently protected underground aquifers. Instead, groundwater is being sold at rock-bottom prices to water-bottling companies or literally given away to oil sands and fracking companies that use copious amounts of water, sand, and chemicals at high pressure to fracture rock so that hydrocarbons can be released

Wildfire is not necessarily bad for the watershed, according to John Moody, a USGS research hydrologist who began examining the relationship between fire

23.3 Mercury and other contaminants released by wildfire enter the food chain, sometimes making large fish dangerous to consume. Photo by Edward Struzik.

and water in 1996 when the Buffalo Creek fire burned 4,690 hectares of forest southwest of Denver. The erosion that often follows, he points out, replenishes the coarse grain sediments that are critical to fish habitat. They also flush in nutrients that are beneficial to those fish.[8]

This was exactly what Silins had been seeing in the Castle Crown Wilderness after it burned in 2003. Algae continued to grow in the cold mountain streams, but not in volumes that starved them of the oxygen that trout and other species need to survive. The fish in burned areas were growing larger than those in unburned watersheds, thanks in large part to the nutrients that continued to be liberated by that fire. Nor were they vulnerable to the potentially lethal warming that can take place when intense fires expose mountain streams to sunlight. The groundwater that flowed out of these mountains was keeping the streams cool enough for trout to survive.

Fish, however, are vulnerable to the chemicals that are often liberated by fire. PhD candidate Erin Kelly discovered this serendipitously in the summer of 2000 when a wildfire in Jasper National Park temporarily shut down a study she was conducting on mercury concentrations in alpine lakes. The 1,100-hectare fire swept through the Moab Lake region, burning three quarters of the lake's catchment area. Rather than giving up, Kelly remained, shifting her study to

the fire and the impacts it had on the lake. The doubling of the lake's nitrogen concentration and quadrupling of the phosphorus concentration was not a big surprise. The nutrients trigged a baby boom of fish. What was unexpected was a five-fold increase of mercury in them.[9]

23.4 Ptarmigan Lake, Jasper. Fire can also release beneficial nutrients into rivers and lakes. Photo by Edward Struzik.

It was fire that liberated the mercury from the soil and from trees in the forest. It was changes in the food web that moved the mercury so quickly up the food chain. In this case, invertebrates feasted on the nutrients and mercury that the fire introduced. Smallish fish like rainbow trout capitalized on that bounty of invertebrates and passed on the mercury to bigger fish like lake trout that prey on them. At the top of this food chain, the concentration of mercury was high enough for government officials to issue a health warning about fish consumption.

Linking wildfire to elevated levels of mercury is not as simple as it may seem to be, however. USGS scientist Mark Brigham was involved in a study of lakes and streams where more than 80 per cent of the freshwater fish had elevated concentrations of mercury. For example, he and his colleagues studied a small lake in Voyageurs National Park that had been directly affected by a wildfire. They found no significant change in levels of inorganic mercury or the more toxic methylmercury in the lake water or in young yellow perch that inhabit the lake. But they did find evidence to suggest mercury was liberated from the soils in the watershed.[10]

The hotter the fire is, according to Brigham, the more the soil burns and the more mercury goes up into the air before settling in another place. This may be good for a local watershed because it removes mercury from the soil in the region. But it isn't good for the ecosystem in which the atmospheric mercury ends up. The Arctic is a case in point. A substantial amount of mercury is carried into the Arctic by air and water currents from human sources at lower latitudes. This has resulted in a ten-fold increase in mercury levels in top predators such as the polar bear in the past 150 years. While polar bears and marine birds can excrete mercury in their hair and feathers, toothed whales such as narwhal and beluga are less able to get rid of the toxin.[11] That's a concern for the Inuit who rely in marine mammals for food.

Then again, it depends on the ecosystem. Bacteria can transform inorganic mercury in an aquatic system into methylmercury, the kind that bioaccumulates, so long as there is plenty of oxygen to energize the bacteria. If not, the inorganic mercury remains relatively stable. The bottom line is that there are too many ecosystem variables in play to suggest that wildfires will lead to an increase in methylmercury in the aquatic food chain. The methylmercury, however, will end up somewhere.

Silins is not alone among his peers in expecting that there will be more surprises in the future as fires burn bigger, hotter, and more often and as climate change, drought, and increasing water consumption adds more stress to diminishing water supplies. Sixteen years ago he was a largely unappreciated forest hydrologist searching, often in vain, for a study site. Nowadays, he and Monica Emelko can't keep up with the requests they've been getting for advice and help. "I don't see it slowing down," he told me. "The impact that wildfire have on watersheds is a challenge that is going to get a lot more intense."

A week after I returned from my field trip with Silins, I reached out again to John Moody, the USGS scientist who studies the relationship between hydrology and wildfire. He told me that there are two things to watch for in the future: the intensity and frequency of fires that burn forests and the extreme precipitation – unleashing a lot of rain in thirty minutes – that follows a fire up to two years after it is extinguished, when soils may not be able to absorb as much moisture as they normally do. Those are the kind of events that can cause severe flooding, extreme sedimentation, and the liberation of undesirable chemicals. "It's expected that fire frequency will increase in many places such as it has in Colorado since 1996," said Moody. "What we don't know is when and where and if we'll see a corresponding increase in the intense precipitation that can result in catastrophic events. Climate in these cases is a factor, as is the geology of the region."

Colorado found that out in the summer of 2020 when drought conditions were as bad as they had been when Hayman burned in 2002 and when the lack of rain coincided with hotter than average temperatures. Records fell everywhere in the west, with the temperature reaching 130 Fahrenheit in Death Valley that August. What made the situation in Colorado even worse was the march of the mountain pine beetle that had killed 7 per cent of the western forest from 1979 to 2012. When the trees die, they no longer absorb the moisture that comes with heavy rain or rapid snow melt. New vegetation eventually balances things out, but not fast enough to absorb moisture in the first few years after the die-off.

Cities such as Denver and Fort Collins were once again scrambling to protect a resource that was essential to several million of its citizens. California found itself in the same situation when there were at one point 650 fires ignited by 12,000 lightning strikes burning across the state. The fire season in California ended with 8,258 fires burning 1.7 million hectares. It was a record year that was almost surpassed in 2021 when 8,367 fires burned nearly 1.3 million hectares. To put that into perspective, the five-year average from 2016 to 2021 was 5,581 fires burning 341,247 hectares.[12] It was one of the reasons why many of the burned-out regions were put on flood alert in October 2021 when a series of atmospheric river storms resulted in flash floods. There was little on the ground to soak up the water that is dropped by these large narrow streams of water vapour that can carry water equivalent to twenty-five Mississippi Rivers.

It's time to connect the dots. In Canada alone, there are twenty-five major watersheds, one of which – the Fraser River watershed – flooded catastrophically in November 2021 when a similar atmospheric river storm swept in from the Pacific. We know little about the flow of these watersheds, or the fish and other aquatic life that dwell in them, because there is, as the World Wildlife Fund pointed out in a comprehensive report published in 2020, no centralized or systematic method in place to monitor these things. What we do know about highly stressed rivers such as those in the South Saskatchewan watershed is that there won't be enough water in them by 2030 to supply the needs for more than half of the communities in the region. Increasingly there have been times, such as 2009, when many others part of the country have faced serious water shortage threats.[13]

How bad can it get?

Sometime soon, both Canada and the United States are going to have another severe, cross-country drought like the one that started in 1999 and ended in 2004. I described it in a report I did for the Munk School of Global Affairs and Public Policy in 2013. At the height of the drought, thirty-two massive dust storms swept across the prairies and Great Plans of the US. Forest fires ignited at five times the ten-year average. Thousands of prairie ponds (or "sloughs," as they are called in the west) dried up, and tens of thousands of waterfowl were unable to find suitable wetlands in which to nest. During the summer of 2001, farmers and towns in the west that drew from surface water in rivers and streams were literally put on rations. On average, they were allocated only 60 per cent of the water they traditionally get.

The drought dried up virtually every part of the country. In 2001, Vancouver recorded the second lowest amount of precipitation since records began a century earlier. Atlantic Canada had its third driest summer ever. For the first time in a quarter-century, farmers across Canada reported negative or zero net incomes. Over 41,000 jobs were lost. The GDP took a $5.8 billion hit.

When another drought like that settles in, there will be less water in our watersheds, more trees killed by mountain pine beetles ready to burn, and quite possibly much more intense fires, because there will be higher temperatures brought on by climate change. It is going to get a lot worse.

24

THE ARCTIC ON FIRE

Along the northeast coast of Somerset Island in Canada's High Arctic, Elwin Bay was carved into a steep, flat-topped mountain range by the glaciers and sea ice that once covered the region. For as long as anyone can remember, hundreds of beluga whales show up every year on an annual migration from Greenland through the Northwest Passage of Canada. Their fidelity to this site is extraordinary, given the fact that nineteenth-century whalers killed more than 10,000 of them – 840 during one notably gruesome, seventeen-day stretch – between 1874 and 1898. Whale oil back then was what fossil fuels are today: a means of heating, lighting, lubricating, and processing textiles and rope.

While helicoptering over the bay with members of the US National Science Foundation–sponsored Northwest Passage Project at the beginning of August in 2019, we saw too many belugas to count in waters riddled with rapidly disintegrating ice. Five hundred? Eight hundred? None of us could guess with certainty. All we knew was that there were likely equal numbers of whales congregating in similar bays and estuaries such as Cunningham Inlet, which we had sailed by a few days earlier.

Polar bears were there as well – a female and cub in this case – homing in on a dead or dying beluga that had presumably swum too far up the estuary before the tide turned and trapped it in water too shallow for it to escape. It happens. Tom Smith, a beluga whale biologist I spent some time with at Cunningham Inlet, saw a polar bear feasting on a beluga while it was still alive.

24.1 Soot from wildfires in the Arctic is blackening glaciers and ice caps and accelerating their meltdown. Photo by Edward Struzik.

Based on what we saw on our eighteen-day, 5,634-nautical-mile icebreaker journey from Greenland through the Northwest Passage of Canada, beluga, narwhal, bowheads, and polar bears appeared to be doing reasonably well. They were there in significant numbers at most of the biological hotspots we visited. But that was not the whole story.

At the US Thule Air Force base on the northwest coast of Greenland, where the expedition began and ended, we basked in shirt-sleeve weather instead of the fog, leaden clouds, and cold rain or snow that are more typical of what to expect in late August. While we were waiting for a military transport plane to take us back to New York, we learned that a month earlier, hikers had to be evacuated from the Arctic Circle Trail because of smoke from a tundra wildfire that was obstructing their ability to see where they were going. The trail stretches up to 200 kilometres from the edge of the ice cap to the fishing town of Sisimiut on the west coast of the country.

Unfathomable as that fire seemed to be at the time, peat-burning tundra fires have been increasing in scale and intensity. Peat is the organic layer of soil that is made up mostly from plant material that doesn't completely decay in places like bogs, fens, and in tundra regions that are typically wet and cool during the short summer growing season. Because much of the lower layers of peat remain frozen in many Arctic and sub-Arctic environments, the carbon stored in that permafrost cannot escape unless the peat is dug up, burned, or exposed to high temperatures over longer periods of time

More than one hundred "intense and long-lived wildfires" had burned above the Arctic Circle during the summer of our voyage, according to the European-based Copernicus Atmosphere Monitoring Service (CAMS). In June alone, these fires emitted as much as fifty megatonnes of carbon dioxide, more than all of Sweden emits in a year. This was as much carbon dioxide as was produced by all the fires that burned in the Arctic from 2010 to 2018. Meanwhile, soot from these fires was falling on and darkening the icecap and glaciers so that they absorbed more heat from the sun.[1]

It is not just soot that is raining down on the Arctic. We know that wildfire emissions and permafrost thawing are contributing to the high levels of mercury that are found in fish, polar bears, beluga whales, and other Arctic animals.[2] We also know that many other toxins found in those plumes of smoke are migrating into the Arctic from fires burning in Eurasia, Africa, and North America.[3] In 2017, Canadian scientists at a High Arctic air-monitoring station on Ellesmere Island, 1,000 kilometres from the North Pole, detected extraordinarily high levels of ammonia, carbon monoxide, hydrogen cyanide, and ethane – chemical compounds consistent with what a pyroCbs event in British Columbia injected into the stratosphere just weeks earlier.[4]

When I talked to the chair of the physics department at the University of Toronto, Kimberly Strong, about that study she co-wrote, she allowed that she was surprised by what she and her colleagues had detected several thousand kilometres from the nearest industrial region. Strong was one of the founding members of the Canadian Network for the Detection of Atmospheric Change, which has established the Polar Environment Atmospheric Research Laboratory (PEARL) at Eureka in the High Arctic. "This was a very unusual event," she told me. "We saw something like this following the massive 2010 fires in Russia, but not anywhere near the concentrations we saw in 2017. Our colleagues in Greenland recorded the same thing happening there."

Because the number of wildfires in the Arctic increases as fast as sea ice is retreating, glaciers are melting, and permafrost is thawing, the resilience of Arctic wildlife is being tested in ways it never has since wild, unsettling swings of cooling and warming followed the hasty retreat of the Last Great Ice Age over the past 12,000 years. No one knows what the thresholds are for extinctions. Mastodons, woolly mammoths, saber-tooth cats, short-faced bears, and eight-foot-long beavers no longer exist because of climate instability in the Arctic. Almost everyone agrees that trouble is not far off for survivors such as beluga whales and polar bears, and for caribou. Caribou are already having a particularly difficult time adjusting to increasing numbers of biting flies and parasites, to overhunting, and to fall and spring rain that turns soft snow into impenetrable ice. And the growing number of Arctic wildfires are rapidly consuming slow-growing lichen, the food caribou most depend on in winter. Because of that disturbance, lichen, grasses, sedges, and other tundra vegetation are giving way to woody shrubs.

It's a little like what's happening in Yellowstone, where the accelerated frequency of fires since the infamous 1988 burn has, in combination with drought and global warming, set the stage for much of the park's lowland forest to turn into grassland by 2050.[5] And it's similar to what is already happening in in the Bob Marshall Wilderness in Montana, where Ponderosa pine are disappearing at low elevations because of drought, high temperatures, and increasing fire frequency.

In some ways, the arrival of fire on the tundra is akin to the human-generated fires that have been ravaging the Amazon rainforest. More than 500 mostly illegal fires consumed as much as 400,000 hectares in the Brazilian Amazon in 2020. Many plants and trees in both places are fire-naïve, meaning they have trouble recovering from fire because, unlike with the evolution of pine and spruce, fire was not something they needed to adapt to or embrace. (Black spruce branches bow to the ground as a way of making it easier for fire on the ground to consume them. The heat from a fire is the only reliable way of releasing the seeds locked tightly up in their cones.)

The most notable difference between places such as Yellowstone or the Amazon and the Arctic is that the Arctic is warming twice as fast or more, and most of its tundra vegetation – lichen, sedges, grasses – is especially vulnerable not just to fire but to higher temperatures and rapidly thawing permafrost.[6] Lichen doesn't necessarily die when it gets too hot and dry. It simply shrivels up and becomes dormant until wetter, cooler weather returns. This process is known as "poikilohydry," one of my favourite words. Nonetheless, when lichen is consumed by fire, it takes a long time for it to re-establish. Reindeer lichens have an initial growth-accumulation period of six to twenty-five years, during which it increases in height every year. It can take up to a century for reindeer lichen to mature in clumps. In places such as California, lichen communities that have been an integral part of the coniferous forest haven't been coming back, even years after fires have been extinguished.[7]

Connecting the dots between wildfire and lichen and declining caribou populations was a long time coming, because most people assumed that the declines resulted from overhunting. But that overlooked the fact that pipelines, predators, disease, fire, and climate were also at play. With scientists Kyle Joly, Randy Myers, and Chuck Racine, Randi Jandt began connecting the dots between climate, fire, caribou, and habitat to see how that played out in tundra areas that burned and those that did not burn in Central Arctic caribou habitat in 1972, 1977, and 1981. What they found was remarkable. A quarter century after those fires, lichen cover on the tundra remained extremely low whether caribou were feeding there or not. Fire, however, was not the only reason for the decline. They found that lichen cover also declined on the tundra areas that did not burn, presumably because it was too hot.[8]

Contrary to what the media tends to report, wildfire has routinely burned the boreal forest of the Arctic. I saw this firsthand in 2004, when both Alaska and the Yukon had their worst fire seasons ever. Seven hundred fires in Alaska burned 6.6 million hectares. An estimated 1.7 million hectares of forest vegetation was consumed by fire in the Yukon, far surpassing the long-term annual average of 120,000 hectares.[9] The fires were burning so big and intensely that summer that I had to cancel plans to canoe the Wind River in the northern part of the territory because the Alaska Highway was periodically closed to traffic. Even if my family and I had successfully made the long drive to the town of Carmacks, where we were to fly to the Wind River by bush plane, the smoke would have likely been too thick for a pilot to safely fly us in and drop us off.

Spectacularly big fires like those that burned in 2004 were viewed more as an anomaly back then, just as the Siberian fires were the year before and as the Black Dragon Fire was when it burned in China and Russia in 1987. But the history of fire in the Arctic is no longer a proxy for the future. As the climate of the Arctic has continued to heat up more rapidly than anyone predicted in the 1980s and 1990s, it was inevitable that the boreal forest would burn as often and persistently as the sea ice was retreating. Record-breaking, or near record-breaking fire seasons in Alaska (2004, 2007, 2015), Siberia (2010, 2012, 2015, 2019, and 2020, 2021), and in the Northwest Territories of Canada (2014) are proof of that. In 2020, an area the size of Florida (60,000 square miles or 15,539,929 Hectares) burned in Siberia, releasing as much carbon as did all of Mexico's fuel consumption, according to the Copernicus Atmospheric Monitoring Service.

The unprecedented fires in the boreal forest of northern Sweden in 2018 required additional firefighting resources from other Scandinavian countries as well as Italy, France, Germany, and Estonia. Rain-soaked, melancholy Swedes were shocked by what had happened that year, as I learned while being interviewed for *Dystopia*, a popular, but very serious, Swedish public radio show that produced a chilling documentary on fire. With the exception of the 2014 fire season that had resulted in one fatality, and 1,000 people and 1,700 animals being evacuated, Scandinavian media were more accustomed to documenting horrors like the wildfires in Portugal that killed sixty-six people in 2017 and those that burned along the coastal areas of Greece the following year. The fires in Greece destroyed two towns, killed 103 people, and hospitalized 140 more. Four thousand homes and 40,000 trees were consumed by the flames.

24.2 Smoke rises from the Bogus Creek Fire, one of two fires that burned in the Yukon Delta National Wildlife Refuge in southwest Alaska on 7 June 2015. Photo by Matt Snyder.

Wildfires bigger than twenty-five hectares were as rare in Sweden as big destructive fires were in California before the Cedar Fire of 2003. From 2008 to 2017, the largest number of annual wildfires that burned more than twenty-five hectares in Sweden was twelve. Part of this has to do with intensive forestry practices and aggressive fire suppression. In 2018, however, there were seventy-four, most of them occurring during the month of July, when it was hotter than it had been for 260 years.[10] Firefighting operations in western Sweden had to be shut down when burning started close an artillery-training range near the Älvdalens forest. Officials feared that the extreme heat might detonate unexploded ordnance. In spite of the large variations in recent fire activity, it is not altogether clear whether Scandinavia is heading towards increased fire danger or a wetter climate that will offset the risk of ignition brought on by climate change.

If there was a tipping point that signalled a new paradigm for wildfire in the circumpolar Arctic, it may have come in 2007, when the Anaktuvuk River tundra fire on the north slope of Alaska accounted for more than half the area that burned in the state that year.

Big tundra fires like the Anaktuvuk had never happened in several thousand years. There was just too much moisture in the frozen ground for a fire to be ignited in the rare event of a lightning strike during summers that tended to be moist and cool. But as temperatures began to rise, that was no longer the case. Since the 1970s, Arctic temperatures have increased by about 2.3 degrees Centigrade. Alaska has warmed more than that, and for every degree of warming, there is a 10 to 12 per cent in increase in lightning. The warmer it gets, the less moisture there is in tundra regions. A lightning strike in such conditions, therefore, has a better chance of lighting up the tundra as it did along the Anaktuvuk River.

Biologist Ben Abbott remembered that fire well when he talked to me about it a few weeks after he returned south. He and some colleagues were at the Toolik Field Station in northern Alaska playing a buggy game of soccer on a gravel helicopter pad when he smelled smoke. Initially he thought nothing of it, because smoke from forest fires farther south occasionally drifted into this part of the world. But he and his colleagues quickly realized that this smoke was drifting from the north slope, where there are no trees. The fire smouldered for a few months before strong, dry winds blowing from the Brooks Range fanned the flames. More than 1,000 square kilometres of tundra burned before October snowfalls quenched the blaze. That fire released as much carbon into the atmosphere as the tundra it burned had accumulated and stored in the previous half-century.

As much as the 2007 Anaktuvuk Fire was a bellwether for what might happen on the tundra in the future, the 2015 fire season was just as remarkable. On 19 June 2015, a slow-moving low-pressure system began making its way through the Alaskan interior. The violent thunderstorms that came with it produced little rain, but they shot out 61,000 bolts of lightning – 15,000 in one remarkably electric day. No one had ever seen anything quite like it, not even in 2004, when 8,500 lightning strikes were recorded in one day during that record-breaking fire season.[11] The 2015 storms triggered 270 fires. Just over two million hectares of forest burned. Seventy homes were lost. Mercifully, no one died.

The only significant difference between the two fire seasons was timing. The 2004 fires burned in July and August. The 2015 fires burned in June and July. Everything else followed a similar pattern. Lightning was the trigger for most of the fires. Heat and tinder-dry conditions fuelled the flames. And most significantly, human-driven climate change was likely the reason for the extraordinary heat, the low humidity, and the violent thunderstorms.

Scott Rupp – a fire ecologist at the University of Alaska Fairbanks and university director of the Interior Department's Alaska Climate Science Center – is certain of that. He and twelve other scientists came to that conclusion after they pored over data that might link the probability of the wildfires of 2015 to anthropogenic climate change. Remove regional warming from the picture, they concluded, and the forests of Alaska would very likely not have burned as big as they did that year. "If the rains hadn't come, I estimate that as much four million hectares would have burned," said Rupp when in the summer of 2016 he took me on a tour of the Agee Creek Fire, which had burned across an oil pipeline a year earlier. "That would have made it the largest runaway wildfire season for a state, province, or territory in modern day North American history." Rupp couldn't resist commenting on the irony of the scene before us that day: scorched earth butting up against a long, giant steel cylinder transporting oil across a previously forested hillside. "The intersection of two worlds passing each other," he said. "An oil pipeline carrying fossil fuels that are warming the climate and a wildfire that is being fuelled by fossil fuels making fire burn more frequently."

The question that Scott and most fire scientists are asking is the one that Californians want answers to after consecutive years of catastrophic fires, including some in 2020 that were ignited by the more than 12,000 bolts of lightning that struck in one week. What's next?

Fire scientist Jill Johnstone tried to answer that question for me later that fall when she took me on a four-day, 1,850-kilometre-long road trip through the Yukon and Alaska, including the crazy town of Chicken, to see first-hand what she'd learned monitoring the plants and trees that grew up after wildfire burned through many parts of the territory and state. "It's like a card game," she said. "After a fire, there is a shuffling of the deck, and once the cards are dealt, the ecosystem has to play the cards it receives. In the boreal, where the ecosystem has very few tree and plant species compared to, say, the Amazon, there are only a certain number of ways that the game can be played out. Controlling forces such as soil acidity, precipitation, heat, permafrost, north and south-facing slopes help determine which plants and trees are most successful. So does climate change, now more than ever before."

There are not one or two but several wild cards in the deck being played in the boreal forest and tundra. The first two are drought and climate change. The third is the cold weather and the permafrost at the northern end of the boreal that makes it difficult for anything but black spruce to push the treeline northward. Stands of black spruce in rapidly thawing permafrost and in areas that have burned resemble a group of drunks exiting a tavern late at night. Because of the thawing and freezing permafrost, the trees have a hard time rooting themselves in a way that allows them to grow straight up. So, they lean against each other precariously.

Shuffle the wild cards into the regular deck and you get a number of "What's next?" scenarios. What if the frequency of fires increases in the northern boreal and the tundra as the temperatures continue to rise? What happens if fire destabilizes the permafrost, as it has done so catastrophically in many places? And what happens if droughts persist, and peatlands ecosystems begin burning extensively, as they did along the Anaktuvuk River?

24.3 Massive slumping is taking place in many parts of the Arctic where fire has burned into the frozen peat. Photo by Eric Miller.

"Scary" is an inappropriate word to use when describing scientific research. But there is no other way of summing up what scientists Merritt Turetsky, Mike Flannigan, Brian Stocks, and Mike Wotton concluded when they assessed what might happen if the peatlands of the boreal forest and Arctic tundra start to burn significantly. Peat fires don't candle as fire does in a coniferous forest. Instead they smoulder, often for months or longer. If they burn down deep enough, they can continue to smoulder beneath the snow through the winter. It's not a stretch to suggest that peatland fires in the Arctic could smoulder for years the way that human-triggered peatland fires in Indonesia have done.

Like everything about this new wildfire paradigm that is unfolding, it's hard to wrap one's head around. There is almost twice as much carbon trapped in permafrost as there is in the atmosphere. As more and more of this carbon is released, temperatures are going to rise even faster. And as more of that black carbon falls onto the glaciers that keep the northern regions cool, the melting and thawing will accelerate.

Having travelled to the Arctic every year for more than forty years, I never return without being surprised by the changes I've seen. The cool spring and summer that followed the eruption of Mt Pinatubo was good for polar bears because it gave them more time to hunt seals on the sea ice. It was a disaster for nesting birds that needed heat to produce healthy chicks. The year 2006 was notable for the polar bear/grizzly bear cross that was shot by an American hunter. In 2012, I was sailing in the High Arctic when one of the most powerful summer storms passed through before petering out somewhere around the North Pole. Being shown Pacific salmon by an Inuit hunter in the eastern Arctic was so surreal that when I told a fisheries expert about it, he insisted initially that I must be mistaken.

What I found most flabbergasting were thunderclouds forming over the sea ice along our journey through the Northwest Passage in 2019. It never produced lightning, as far as I could tell, but a thunderstorm that formed much farther north, just 500 kilometres from the North Pole that same week, shot out forty-eight bolts of lightning.

Given what happened in the Arctic in 2019, it was not entirely a surprise to see Siberia light up again the following year when new records were being broken. For the first time in recorded history, Siberia baked in temperatures exceeding 100 degrees Fahrenheit. Those of us who tracked the conflagrations experienced a sense of déjà vu. In June and July, fires released more carbon than all that is emitted annually Denmark, Sweden, Norway, and Finland combined. In some places, 15,000 years of peat accumulation went up in smoke that summer.

There were no words to describe what happened in Russia in 2021 other than to note that the 18.16 million hectares that burned made it the largest in recorded history.

25

THE BIG SMOKE

During the first week of August of 2021, a vast, thick, and very acrid blanket of cloud formed by hundreds of fires burning in the boreal forests of northern Russia drifted more than 3,000 kilometres to the North Pole. It appeared to be a first in recorded history. The Republic of Yakutia, officially known as the Republic of Sakha, encompasses a little over three million square kilometres in the Far Eastern Federal District. It is also known as "the Kingdom of Permafrost," and the coldest place on Earth. That summer, it turned into the "Capital of Wildfire."[1]

Long-range transportation of smoke from wildfires is nothing new. The Chinchaga Fire of 1950, which burned for 222 days in northern Alberta, created what was then the world's largest blanket of smoke. The dark terrifying mass of cloud drifted southeast into Ontario and New York State, turning noon into night, grounding planes, and forcing cities and baseball parks to turn lights on mid-afternoon. When the pall eventually made its way across the ocean to Europe, it resulted in a blue moon that led to panic and a run on some banks.

In 2018, smoke from five pyroCbs that erupted in British Columbia was detected at research stations in Greenland and on northern Ellesmere Island in the Canadian Arctic. Two years later, scientists watched with awe as a pall of aerosols from the Australian bushfires circled the southern hemisphere twice, sending as much smoke into the stratosphere as one would expect from a volcano or nuclear blast. The haze was thick and widespread enough to cool the globe for a time.[2]

25.1 The 2014 fires in the Northwest Territories produced a "summer of smoke" and a doubling of emergency hospital visits by people seeking treatment for asthma.

The fact that smoke can travel such long distances is nothing new. I was made aware of that in the summer of 1995 when I was canoeing the Coppermine River, which flows through the boreal forest of the sub-Arctic into the tundra before spilling into the Arctic Ocean. Some 8,500 wildfires burned more than seven million hectares across Canada that year, making it the second biggest fire season since national records were logged. Only 1989 produced more fires and more area burned. My partner and I were more than a thousand kilometres from the nearest fire. But there wasn't a day in the three weeks of paddling when we didn't smell smoke or see the sun glowing orange through the haze.

And yet, the realization that smoke from wildfires might have an impact on public health has been as slow in coming as was acceptance of the dangers associated with the inhalation of tobacco smoke. Maybe it was because people didn't want to believe that air that smelled like a campfire on a summer night could do them harm. It was smog from automobile and industrial smokestacks that worried them in the latter part of the last century.

Public health officials began to realize there was a problem when a link was established between smoke from a fire in the province of Quebec in 2002 and a mysterious, troublesome, thirtyfold increase in air pollution downwind in Maryland. The research highlighted the need for studies that assess the impact

that fires, both distant and local, have on public health. Linking wildfire smoke to adverse health effects is difficult because smoke often lingers for only a few days. It is also unethical for researchers to conduct experiments that expose people to anything that is likely to harm them.

25.2 In 2008, smoke from a peat fire in the Pocosin Lakes National Wildlife Refuge in North Carolina resulted in a spike in hospital admissions in forty counties.

The answers that are beginning to come, however, are alarming. When a peat fire in the Pocosin Lakes National Wildlife Refuge in North Carolina produced smoke and haze intermittently for several weeks in 2008, hospitals serving forty surrounding counties recorded a dramatic spike in admissions for asthma and congestive heart failure. A team of epidemiologists found a clear correlation between the smoke from the peat fires and those suffering ill effects.[3]

The death and hospitalization count has risen dramatically since then, largely because public health officials are now paying attention, and because fires are burning bigger and hotter, and for much longer. Stanford University researchers believe that in California alone, between 1,200 and 3,000 people died in 2020 as a result of inhalation of smoke from wildfires. "These are hidden deaths," Marshall Burke, an associate professor at Stanford who calculated the death toll, told the *Mercury News*. "These are people who were probably already sick but for whom air pollution made them even sicker."[4]

The concerns do not stop there.

Exposure to wildfire smoke may also be associated with thousands of additional coronavirus infections as well as hundreds of deaths during the wildfire season of 2020.[5] It is also likely having an impact on children, who are just as vulnerable to smoke as are the elderly, and on those who suffer from respiratory problems. The single-day North American record for school closures due to smoke was 15 November 2018, when more than a million children in California were told to stay home.[6] The situation is even worse in Indonesia, where tens of thousands have died as a result of smoke from fires that have flared and smouldered uncontrollably as peatlands were drained and burned to make way for palm plantations. A team of British scientists concluded that by eliminating fire in the region, 59,000 deaths could be averted annually.[7]

Much of the concern about smoke has focused on particulate matter: microscopic particles that can penetrate deep into the lungs. If these particles are large enough, as they often are in wood smoke, they can cause a wide range of health problems, from burning eyes and a runny nose to aggravated chronic heart and lung diseases. Exposure to particle pollution is also linked to premature death. What's missing in the data is the effect of the chemicals in smoke. Benzine and acrolein, commonly found in tobacco smoke, are also found in smoke from wildfires. Fires also release ozone and carbon monoxide. Concentrations of these gases produced by a wildfire may become high enough to affect the health of firefighters on the ground, but rarely high enough to affect people living in or near a forested area that is burning. But when ozone and carbon monoxide are transported hundreds and sometimes thousands of kilometres to cities that already have high levels of pollution, they can exacerbate health dangers.

That is what happened in Houston in the summer of 2004, when smoke from fires in Alaska, the Yukon, and western Canada migrated south. The ozone concentrations in the air over Houston – which were already high – rose by 50 to 100 per cent from pre-existing levels. The quality of the air was so bad in Harris County, which at the time ranked third in the country in toxic emissions, that it was almost off the charts.[8] The challenge in accurately measuring ozone is that high concentrations of volatile organic compounds emitted during the burning of trees interfere with detection instruments. Therefore, the impact may not have been as severe as previously thought, or it could have been worse.[9]

Oxygenated organics such as formaldehyde and acetaldehydes, both of which are carcinogenic, are found in smoke as well. Hotter fires tend to result in more free radicals – unstable atoms that can damage cells – being released into the atmosphere. Free radicals are likely having an effect on human health, but no one knows with any degree of certainty how much.

Then there's mercury, perhaps the most dangerous ingredient in smoke because it has the longest-term environmental impact. Much of the mercury in the environment comes from decades of burning coal and from other industrial emissions that settled into the soil and were absorbed by plants and trees. When a wildfire sweeps through, it liberates mercury in the same way that it liberates arsenic and other chemicals. Those plumes of smoke from wildfires in Quebec that hospitalized so many people in the northeastern part of the United States in the summer of 2002 contained a significant amount of mercury.

Even though soil-bound mercury volatilizes during fires, it doesn't elevate atmospheric mercury to such a great extent that it's a health concern. But this mercury can, over time, fall back to Earth's surface, where it builds up in the food chain. In Voyageurs National Park in northern Minnesota, some fish have concentrations of mercury ten times higher than is safe for people to consume. This mercury in the aquatic environment is being passed on to loons and bald eaglets and eagles that eat fish. It's a concern everywhere in the upper Midwest, but not one that is affecting the animals' reproduction.[10] Impairment of reproduction occurs when this neurotoxin crosses the blood-brain and placental barriers, allowing it to react directly with the brain and fetal cells.

Alarming as this is for the futures of both public health and aquatic ecosystems, it's what we don't know about wildfire smoke that's perhaps an even bigger concern. Residents of the small mountain town of Libby, Montana, found this out shortly after the Environmental Protection Agency came in 1999 to clean up asbestos from a mine that had been operating in the Kootenai Forest for approximately eighty years. What was expected to be a months-long mop-up ended up being declared "an imminent and substantial endangerment."[11] The mine was added to the National Priorities List of Superfund sites when a study showed that the asbestos in Libby's vermiculite is a unique mixture of at least five chemically similar fibers. This form of asbestos tends to fracture, forming long, needle-like fibers that stab the lungs. As the fibers scar the lining of the lungs, they harden over time, making it difficult for people to breathe on their own. Workers involved in processing Libby vermiculite also demonstrated an increased risk for mesothelioma, an extremely aggressive form of cancer that often goes undetected until it is at an advanced stage.

The EPA was initially so busy trying to decontaminate homes, businesses, playgrounds, and schools that were contaminated with asbestos that it didn't have the time or the inclination to look for fibers farther afield. That was taken on by Tony Ward of the School of Public and Community Health Sciences at the University of Missoula in Montana and by Julie Hart and Terry Spear at Montana Tech. When they and their colleagues began to assess whether logging could be safely conducted in forested areas near the old vermiculite mine, they were surprised by what they found. After first analyzing soil and tree bark collected in close proximity to the mine, they discovered that asbestos fibers were imbedded in the bark of trees far from the site. "That was a game changer," Ward told me in an interview. "From that point on, our research went in another direction. We decided to go back the next year to see if this was occurring in areas outside of the mine site, within the town of Libby, and in railroad corridors outside of Libby. The results of those investigations demonstrated that it was."

The potential for these fibers to be liberated by wood harvesting, woodstove burning, and wildfire was tested by Ward and his colleagues, in fire chambers by the EPA, and in small experimental burns conducted by the US Forest Service at the mine site. The findings were alarming enough to compel the EPA to declare a public health emergency in 2009. It was the first time in its history that the agency had gone that far. The reasons were clear. Because asbestos is resistant to fire, the almost indestructible fibres attach to wood that is burned. Homeowners who used this contaminated wood for heating could easily have reintroduced asbestos into their houses.

One of the more sobering insights came from Nikia Hernandez of the US Forest Service in July 2016. At the time, Hernandez had twenty-three years of experience fighting more than a hundred wildfires. In an email to the EPA's supervisor of the cleanup site, he summed up what scores of scientists have learned about fire behaviour: that it can create its own weather and carry ash and other contaminants long distances before they fall to the ground. By way of example, he described how ash from the 5,000-acre Klatawa Fire that burned ten kilometres southwest of Libby in August 2015 forced the evacuation of some people in Libby. The ash ended up on cars in town, including Hernandez's vehicle, which was sixteen kilometres away from the edge of the fire.

"My point here is just to make sure everyone understands that wildland fires have enough energy in them to loft and transport particulates of very small sizes, ash, and even larger pieces of debris," he wrote. "How far all these things are transported is dependent on many factors such as size and shape of the particle/debris, size and height of the convection column, topography, and transport winds to name a few."[12]

Hernandez also noted something else that most wildfire scientists understand. Despite the great strides that have been made in predicting fire behaviour, big wildfires are inherently unpredictable. If smoke from fires in Yakutia can travel 3,000 kilometres to the North Pole, then it's fair to say that a fire burning in the asbestos forest of Libby is not just a local problem.

Nor was the contamination in Libby an isolated incident. Toxic fallout from wildfires has happened elsewhere. In May of 2003, an experimental burn in the forest surrounding the Chernobyl nuclear site released cesium-137, strontium-90, and plutonium-238, -239, and -240. "That was crazy," said Sergiy Zibtsev, a forestry professor at the National University of Life and Environmental Sciences of Ukraine. "That place was really contaminated."[13] Thirty-five years after the explosion, officials were just beginning to use computer models to explore a wide range of fire management scenarios in the Chernobyl region.[14]

25.3 Firefighters are especially vulnerable to the dangers associated with smoke inhalation. The Pioneer Fire in Idaho in 2021. Photo by Kari Greer.

It's a lesson for caretakers of thousands of abandoned mines sites located in the forests of North America. Arsenic that spewed out of the Giant Mine site in Yellowknife in the Northwest Territories for decades ended up in forests and adjoining peatlands, some of which narrowly escaped an intense fire that burned in the area in 2014. The abandoned Clinton Creek asbestos mine in the Yukon is so unstable that government officials warn people to keep out. Although it was shuttered in the late 1970s, there are no plans to begin cleaning up the site until sometime in 2026. Once that begins, it will take at least fifteen years and a half-billion dollars to finish the job.

If researchers at Scripps Institution of Oceanography at University of California in San Diego are right, we may well be underestimating the toxicity of smoke from wildfire. After examining fourteen years of hospital admissions data, they concluded that the fine particles in wildfire smoke can be several times more harmful to human respiratory health than particulate matter from other sources such as car exhaust. While this distinction has been previously identified in laboratory experiments, the new study confirms it at the population level.[15]

What's becoming increasingly clear is that the health effects associated with wildfire smoke are costing states, provinces, and territories more than they pay to suppress fires. That's what happened in 2008, when fire in the Pocosin Lakes National Wildlife Refuge in North Carolina burned 18,200 hectares of forest and peatland over a two-hundred-day period. The $48 million it cost to treat people suffering from respiratory and cardiac problems was twice the amount that it cost the state to put out the fire.

Many firefighters have accepted the possibility that they might not live as long as they would were they in another profession. "It is what it is," said Jamie Coutts, a fire chief who fought two notorious wildfires, one in his hometown of Slave Lake and the other in Fort McMurray. "The chances of me dying of cancer are probably a lot higher than they are for most people. I was diagnosed with asthma and later with lung sarcoidosis. I can't say for sure that smoke was the cause, but I suspect that it was. My message to the public is that if it can harm me, a guy who entered the field in good health and very good shape, it can harm you."

26

FIRE NEWS

In November 1871, the *New York World* published a first-hand account of the Peshtigo Fire that set a standard for the role that newspapers and magazines would play in shaping policy and public understanding of wildfire over the course of the next 150 years. The article was written not by a journalist but by Wisconsin farmer Lucile Mechand, who recounted how, in the hours before fire came to her property, she had been sitting down to have breakfast with her husband, her mother, and her six-year-old son, Louis. There was an unexpected knock on the door and a neighbour asked Mr Mechand to come with him. Her husband returned sometime later with the news that the forest was on fire a few miles away, and that gale force winds were pushing it towards them with "frightful" speed.

Lucile wasn't surprised. Fires had been burning in the area for days. Smoke was everywhere. Nonetheless, the family decided to stay where they were. The air, however, became so "sultry" that they had a hard time breathing. Still, no one panicked until Lucile looked at her mother sitting in the corner and noticed that "she seemed restless and that her eyes shone with a light such as I have sometimes seen in the eyes of a wild beast."

Then there was another knock on the door. This time, a burning branch of pine fell at Mr Mechand's feet as he greeted the visitor. Smoke poured into the house, and he ordered his family out as the wood-framed structure began to go up in flames. Lucile's mother hesitated, darted to the cupboard, and tucked something inside her shirt before being dragged along.

As they were running through the forest, her mother broke away. Lucile gave chase but lost sight of her and got lost in the woods. She eventually found her son but saw no sign of her husband. Night fell, but not in the typical manner.

"On ordinary nights, it should have been dark, but there was a nameless glare. Yet not a glare, but a horrible renet which came down from the sky and mingled with the smoke. Hardly had I risen from the ground when in the direction of the woods on the other side of the clearing, I heard a clashing noise, a mingled gnashing and hoarse barking, which I instantly recognized as wolves, and I had scarcely time to snatch up Louis, and run behind a magnificent pine tree, whose trunk was at least two metres in diameter, before I heard them scrambling up the side of the other hill and felt them rush by me. I looked out and could see their eyes coming towards me like the wind. They did not stop for an instant, and when they passed, there came in their tracks a herd of deer, uttering cries that seemed almost human in their intense agony. They ran blindly, for something more terrible than wolves was behind them; they struck the tree and were hurled back by the shock of them falling back on those below.

"The stampede seemed to last for a full ten minutes, and when it was over and I trembling with fear, dared once more to emerge from my refuge and look across the clearing, I saw the woods at its edge already burning – saw it lurid through the smoke, and felt its terrible heat on my face. I turned and fled in the wake of the deer and the wolves."

The heels of Lucile's shoes were stripped bare. Her feet and ankles were bloody. After carrying her six-year-old son in her arms for miles, she eventually fainted from exhaustion. She knew nothing of what had passed until someone shook her from her slumber. It was her mother, brandishing the knife she had taken from the cupboard, "eyes blazing with that hated light of insanity.

"Fine night this! Night – it is day! Look at the red light – 'tis the light of dawn. Le jour! Le jour du jugement est arrivé! And the rocks are burning. Call upon them to fall on you. The clouds of thunder and the day of doom. The Lord is coming and the wheels of his chariot burn with his might driving. Let us go to meet him in mid-air."

Lucile grabbed her terrified son and asked if her mother planned to stab her with a knife if she didn't sacrifice herself to the flames as she was told to.

"Kill you! Yes! Why wait," her mother replied. "The Lord calls and the devil drives. He has let loose his imps against the world. The trees fall crashing in the forest; for all hell's demons pull them down with hooks of fire ... I will kill you: would you burn to death? You should go up – up higher than the moon and beyond the fire. Come, let us go."

Lucile ended up overpowering her mother. She was determined to strangle her until she came to her senses and collapsed. When she awoke, three men

26.1 Forest fire in America as depicted by *The Illustrated London News.*

were hovering over them. They had come to the rescue. Lucile, her mother, and her son recovered after being taken to Green Bay.

The story ends with Lucile lamenting that she has heard nothing from her husband.

It's possible that this story was ghost-written and embellished for effect. Mark Twain, after all, was guilty of embellishing a story about a fire that he supposedly started in the Tahoe Basin of Nevada in 1861. It would have been exceptional for a farmer like Lucile to write something with such skill and drama. But a French-Canadian surname like Mechand suits the locality where there are French-Canadian villages such as Lac qui Parle, Marquette, Nicollet, Eau Claire, Prairie du Chien, LaSalle, Little Canada, Faribault, Bottineau Blvd., Centreville, Lac Courte Oreilles, Duluth, Radisson, Roseau, and Baudette.

And the description of the fire – the smoke, the "nameless glare," and the wolves and deer stampeding blindly through the woods – is spot on. Only a person who knew something about fire and life in the forest is likely to have written it. Maybe it was a ghost writer. But no one in Chicago, Toronto, New York, and other cities where the article was published could read it and not be struck by the fact that wildfire was a terrible demon lurking in the woods – a warning from God perhaps, as the mother thought, or the result of carelessness by people living and working in the forest.

The Peshtigo Fire, and the many others that burned in North America in the late nineteenth century, did not persuade decision-makers to act on the problem as the Big Burn had in 1910. But articles like these set the stage for wildfire to become a staple in serious newspaper coverage – coverage that eventually influenced politicians and the public to accept that fires needed to be suppressed by whatever means necessary. The media was onside much earlier than the majority of lawmakers.

Destructive fires in the nineteenth century came at a perfect time for newspaper editors. The competition for readers had become fierce thanks to growing literacy, a rapidly rising population, and most of all, to technology. Prior to 1847, when telegraphic news began linking Europe to North America, current news was local or regional in nature. It took up to ten days or more to get news delivered to the door or to the club by postal services. News of some fires such as the Dark Day of 1780 often came through personal letters. By the time information about a fire like Miramichi reached New York and crossed the Atlantic in 1825, it was old news that was not worthy of further pursuit because there was no way of interviewing the victims or talking to officials like Governor Howard Douglas, who rode 200 kilometres on horseback to be among those left homeless.

News began to travel with more ease in 1849, when the Harbor News Association opened the first news bureau outside of the United States in Halifax, Nova Scotia, to meet ships from Europe before they sailed on to New York. In 1856, the Associated Press set up bureaus in Washington, DC, and New York. Then, in the 1870s, the Associated Press, Reuters in England, Havas Agence in France, and Wolffsche Telegraphenbüro in Germany began sharing services, making it easier to distribute stories internationally. Improvements to the rotary press allowed papers like the *New York Tribune* to print 18,000 papers per hour in the 1870s. It was big business and very profitable.

For editors, destructive fires stood out from other natural disasters such as hurricanes, tornadoes, and landslides because they were a relatively new phenomenon and they often lasted for more than a day. They seemed to flare up everywhere with mysterious frequency and brute force. In the nineteenth century, most scientists still believed in *abiogenesis*, the theory that trees grow back as a result of spontaneous generation. Plants simply came into being magically, if inexplicably, animated by some catalyst either divine or chemical. Fire was the great destroyer, not a birther. Only God, or something in the soil, could trigger regeneration after a fire.

The story of wildfire not only begged to be told, it had to be told well to stand out from the competition. You could see how this played out in the *New York Times* in 1870. Its coverage of the Ottawa Valley fires was as remarkably graphic and as well told as Lucile Mechand's narrative above. In one of the stories, four American entrepreneurs narrowly escape death while travelling through the region. Their horses "trembled," and their "eyes ached." Shooting embers set their clothing on fire as the "night advanced and the glow of the fire got brighter." Outside a farmhouse, they find the body of a man clasping a young girl. Closer to the house, a badly burned woman is holding a dead baby. At another farmhouse, they listen to a man and woman tell the heartbreaking story of trying to get all their children out of their burning home. The seven-year-old girl, "the pet of the family," insists that the father take the younger ones first. In the confusion, the father assumes she has managed to get out on her own until he hears her cry, "Oh! Papa, take me, take me too." The father rushes to the house, mortified to see the building collapsing down on her.[1]

Fire was not just a killer in those days, it was also a business story. Reporting on fires in the New Jersey Pinelands and in western Massachusetts and Tennessee in 1874, the *New York Times* noted this: "The fog and smoke on the Mississippi rivers were so dense yesterday as to force the tying up of steamboats, and for the first time completely blocking the wheels of commerce."[2] Over time, accounting for the economic fallout came into play with the help of insurance companies, many of which refused to insure properties considered to be fire hazards. Typical of this kind of coverage were those in 1880 which described how fires consumed an estimated fifty sawmills, hundreds of horses and cattle, seven hundred homes, and cranberry marshes that had dried up after no rain had fallen for weeks.[3] Dollar figures were often added to highlight how destructive conflagrations such as these were.

Between 1880 and 1890, the *Times* published close to 200 stories on the economic toll wrought by fires across the US and Canada. Something needed to be done, the paper proclaimed in 1889. "The movement for the preservation of forests that was begun three or four years ago has had very useful results. It has impressed people with the conviction that forests will not maintain themselves if anybody who gets possession of them is permitted to clear off the lumber with no other thought than that of immediate profit. It has for the first time brought home to the public the fact that our heritage of forest is not only exhaustible, but is in the way to be rapidly exhausted unless something is done to regulate and to repair the ravages of fires and of lumbermen."[4]

Newspaper coverage has proved invaluable from a historical point of view. Prior to the turn of the last century, no government kept records of fires. We only know that big fires burned regularly because of newspaper coverage and local history books that had very limited distribution. When Pennsylvania burned again in 1879, the *Times* recalled how just a few years earlier, the state had been visited by "the most destructive fire ever known." Not only was valuable timber and property lost, the reporter lamented, so too were "the eggs of pheasants and other birds, for which this region is noted."[5] And when the New Jersey Pinelands lit up in April of 1880 with "one of the most destructive forest fires that have ever visited the region," it was acknowledged that this had happened before.

Each year brought fires that were noteworthy for their destruction. In 1883, it was Bar Harbor in Maine, where the black smoke, "like a heavy pall, envelopes the island and renders respiration difficult."[6] In 1886, it was Wisconsin burning again. Yellowstone flared up that year as well, "beyond human control." The cavalry was brought in to fight the fires.[7] Minnesota burned in 1889. In 1892, it was Connecticut, and then the New Jersey Pinelands setting yet another record in 1895 with a fire that the *Times* described as "the greatest fire ever known in the South New Jersey woods."[8] By 1896, almost everyone who could read – and the majority of people in North America were literate – knew about this dragon in the forest.

As the *Times* often pointed out, no one in government had a clear plan that everyone could agree on. With an election looming in the fall of 1896, Interior Secretary Hoke Smith called on the Academy of Sciences to establish a commission to create a national forestry policy. Gifford Pinchot, the youngest appointee, was the only member of that commission who was a forester. The commission's findings, which were widely reported by the media, set the stage for the creation of thirteen forest reserves covering 8.5 million hectares, the establishment of Mount Rainier National Park, and the employment of seasonal rangers to patrol the reserves and fight fires.

For Pinchot, this was not enough. He lamented the fact that the Department of the Interior, "with its tradition of political toad-eating and executive incompetence," was in control of the forests." They were, he said, "incapable of employing the powers," of the Organic Act under which the forest reserves were established.[9] One of the first things Pinchot did as head of the US Forest Service was to hire individuals whose sole job was to scan through newspapers to get a sense of where and how often fires burned. He used the data not only to convince politicians that something needed to be done; he used the data to get the media and the public on side.

It was remarkable how much newspaper coverage Pinchot got. It didn't hurt that he was handsome, athletic, rich, and unmarried during his tenure as the US Forest Service chief. This alone was wonderful fodder for gossipy papers like the *Washington Star* that noted with some tsk-tsking that Pinchot "cares nothing for women," even though in private life he clung to the idea that Laura Houghteling, the woman he lived with and was about to marry before she died of tuberculosis in 1894, was always near or present, reading books with him and advising on matters of public policy. His friendship with conservationist John Muir certainly didn't hurt either. Muir was a wildly popular figure, a prolific writer, founder of the Sierra Club, and someone worthy enough to be Roosevelt's camping partner for three nights in Yosemite.

Pinchot attracted the attention of serious journalists with well-worded press releases, government advisories, pamphlets, and other reports. There were 10.8 million of them distributed from 1898 to 1909. It was, according to journalism professor Stephen Ponder, a "benchmark of an important historic development in the role of the executive branch of government in leading public opinion."[10] Pinchot understood that "nothing permanent can be accomplished in this country unless it is backed by sound public sentiment. The greater part of our work, therefore, has consisted in arousing a general interest in practical forestry throughout the country and in gradually changing public sentiment toward a more conservative treatment of forest lands."[11]

26.2 Gifford Pinchot was often pilloried or praised in the media. Cartoon by Udo J. Keppler in *Punch*, 15 September 1909.

To help get the message out, Pinchot recruited men with literary skills and journalism backgrounds to staff one of the government's first press bureaus. He also persuaded newspaper publishers like William Hutchinson Cowles, a Yale classmate, to print favourable front-page editorials on forestry.[12]

While Pinchot served as US Forest Service chief between 1905 and 1910, his name appeared in *New York Times* and *Los Angeles Times* headlines on more than one hundred occasions. The *Washington Post* had him in their banner headlines more times than that. Cartoonists couldn't get enough of him. Pinchot was a media darling in the east. "A Highly Cultured Dreamer" was the way one headline in the *Times* put it. "Mr. Pinchot's Fight" and "Mr. Pinchot's Mission," stated two others. "Savior of the National Forest" was how he was referred to in more than one article. The most flattering came from the *Boston Globe* when it suggested that "theoretically, he is a Republican, but practically is Gifford Pinchot, uncompromising idealist and implacable foe of those interests which he considers inimical to the public welfare. He never compromises and he never backs up."[13]

It wasn't just big city newspapers that reported on his career with admiration. In 1906, *The West Point Times Herald* in Mississippi was so smitten with Pinchot that it suggested in an editorial that Pinchot would make a perfect president. "He is a man of great wealth and high ideals and gives his entire salary to charitable purposes. If Mr. Pinchot's plans are carried out, the United States will be rich in timber wealth … it is believed that he is fast developing into strong Presidential timber. If a candidate were selected purely from patriotic grounds, Mr. Pinchot would receive almost the unanimous vote of the country."[14]

Newspapers out west were not quite as respectful because ranchers, timbermen, and miners resented the federal government's takeover of so much valuable lumber in the region. *The Argonaut* of San Francisco correctly predicted Pinchot's fall from power when he got into a tangle with Ballinger over coal in 1909. "Pinchot will either retire or he will sink back to his normal place as a subordinate functionary in the Department of Agriculture. He will be what nature and fortune designed him to be, what he was before Roosevelt 'found'

Puck
BER GRAB
LAND GRAB
GREED

him, a pink tea hero [one who speaks to frivolous gatherings of mostly women]."[15] And following Pinchot's departure, the *L.A. Times* facetiously noted that the verdict on his tenure was that he was a "lamented manufacturer of vacuums."[16]

Unkind as some of the coverage was in the west, there was, as P.T. Barnum famously stated, "no such thing as bad publicity." Pinchot, thick-skinned, knew this to be true. He was too good at framing an issue at the right time for newspapers to ignore him. And some of the press, including muckraker George Kibbe Turner, knew they were being used. In a 1912 article in *McClure's Magazine*, Turner described Pinchot and Roosevelt as the two most talented men in American politics.[17]

Even after he had been unceremoniously fired from his job as chief of the US Forest Service, Pinchot's opinion still packed a punch when the Big Burn of 1910 caused so much death and destruction. In a carefully timed and worded press release, Pinchot blamed Congress, and Senator Weldon Heyburn of Idaho in particular, for refusing to provide adequate funding for the Forest Service. "If even a small fraction of the loss from the present fires had been expended in additional patrol and preventative equipment some, or perhaps nearly all, of the loss could have been avoided."[18]

Pinchot used the media to demonize fire as a force that unnecessarily caused so much "loss, suffering and death."[19] Readers were able to appreciate what he was saying because the news had been constantly reminding them of the tragedies that often came with fire. They weren't interested in George Hoxie, the California timberman who warned that "the prevention of fire may be made so complete as to menace the forests with greater danger than they now incur." Much has been made of what Hoxie and others thought about prescribed burning at the time, but *Sunset*, the magazine Hoxie wrote the article for, was not the *New York Times* or *L.A. Times* or the *Washington* Post. It was a promotional tool for the South Pacific Railroad, which ran a rail line between New Orleans and San Francisco. The partners in the railroad had an iron grip on politics in the Sunshine State, but not in the rest of the country. By contrast, Pinchot's command over the media made it easier for Henry Graves and future forest chiefs to get what they wanted. What they wanted was to remove fire from the landscape in every way possible, no matter how ill-founded that idea later proved to be. Newspapers helped them do that.

In Canada, the media was later to champion the wildfire challenge, the Miramichi (1825), Ottawa Valley (1870), and Saguenay fires (1870) being short-lived exceptions. The reason was that newspapers such as *The Globe*, the *Toronto Daily Star*, and the *Montreal Star* tended to align themselves more closely to political parties and to corporate entities such as the railroad and pulp and paper companies than many of the bigger American papers did. Some papers were owned by cabinet ministers such as Clifford Sifton, who was publisher of the

Manitoba Free Press, and Frank Oliver, publisher of the *Edmonton Bulletin*. There was no room for "muckraking journalism" as there was south of the border, where journalist such as Frank Norris, Lincoln Steffens, Ida Tarbell, Upton Sinclair, and Jacob Riis took on corporate power, big trusts, and corrupt governments.

Serious, unbiased coverage of wildfire didn't begin in an impactful way until 1911, when the Porcupine Lake area near Timmins in northern Ontario burned. It was Canada's Peshtigo. At least seventy-three people, likely more, died. More than 200,000 hectares of forest burned following two months of small fires smouldering in the region. The day after the big conflagration, the *Toronto Daily Star* devoted half its front page and a full inside page to the story. *The Globe* did the same the next day. Both papers were on top of the story for weeks, with headlines such as "Towering Wall of Flame Jumped Half-Mile Lake,"[20] "Dug Little Lake and Saved 47 men," and "Cochrane Dismisses a Chief Drunk During Fire." (The chief in this case got blind drunk in what ended up being a shootout.)

This wasn't exactly "muck-raking journalism." But neither was it the staid coverage that typified newspaper reports up until that time, or the "journalistic trope" that fire historian Stephen J. Pyne referred to when he noted how Canadian fire scientist James Wright mocked the media, suggesting that journalists still believed that fires "just happen like Spanish influenza."[21] There is nothing in newspaper archives to indicate that coverage in the major papers was quite as vapid as that. In the aftermath of the Porcupine Fire, *The Globe* stated insightfully that "Ontario's most direful conflagration … should shock the province into demanding adequate means of protection." The paper acknowledged that drought was one reason the fire was so destructive. But it placed most of the blame for ignition on prospectors, "who are inclined to regard fire with indifference knowing that it facilitates their work" and settlers who "are unpardonably careless in starting fire to clear their land." Railroad construction and road builders were not let off the hook. Nor were the pulpwood companies that left slash that "becomes most explosive in dry weather."

"And against all of this," *The Globe* added, "we have the old inadequate staff of largely inexperienced men trying to beat out each incipient fires with blankets and balsam brooms. The inevitable result is lamentably with us. The changed conditions should be met with adequate protective measures … it is a time for rising to the demands of a new situation."[22]

Newspapers invested a lot in the coverage of this wildfire. The *Toronto Daily Star* sent a correspondent to the region to describe in detail how men in nine canoes were still dragging the lake searching for bodies days after the fire was extinguished, how women stood nervously in a theatre waiting to hear if the names of their missing husbands would be announced in the list of dead, and how a piano player entertained the destitute in a hotel. It published a sorrow-filled story devoted to a funeral held at Deadman's Point, which had been named a year earlier in an honour of a Frenchman who died of fever. "It was little thought of at the time that before many months, this place would be turned into a regular cemetery for the dead of the new land," the reporter wrote.[23] And there was the "indescribably awful" scene at the wharf where men, women, and children could not resist gazing at the bodies that had been lined up, side by side, waiting to be transported out. It took fourteen men to lift the coffin of Robert Weiss, "the giant manager" of one of the mines, who was reported to weigh 440 pounds.

Because of this coverage, fires were no longer "fireworks spectacles," as Elizabeth Simcoe, the wife of the lieutenant governor of Upper Canada, described them in her diary many years earlier. One could not expect to walk through a forest fire, as she claimed to have done, without suffering miserably. There was, however, one big thing that separated Canadian reports on wildfire from those of American newspapers. Until the 1920s, there was no one in Canada like Pinchot who the media could rely on to articulately and authoritatively make sense of fire and suggest what needed to be done to prevent future catastrophes. Nor was there a political leader like Roosevelt, who had the fortitude to stare down lumbermen like F.E. Weyerhaeuser, when he accused him and the others in 1905 of being "skinners of the soil" and "despoilers of the national heritage."[24]

Bernhard Fernow, the so-called "Father of Forestry" in America and the man who the *New York Times* said "made Pinchot,"[25] did not reach out to the media when he was made the dean of Forestry at the University of Toronto in 1907. Clifford Sifton, the head of the Commission of Conservation, was a close as Canada got to Pinchot, which is not saying much. Up until the Porcupine Fire, Sifton was making the case in the media for reforms and legislation that would require railway companies to maintain efficient fire protection along their lines. When that legislation was finally and grudgingly passed, he expected the provincial governments to appoint special inspectors to see that the law was carried out.[26] This still didn't happen, at least not with the efficiency or resolve he had expected.

The Porcupine Fire should have been a golden opportunity for Sifton to pounce on that and many other oversights. But when asked to comment on the fire and the devastation it caused, he declined to say anything until an inquiry was called. So it was left to the media, once again, to drive home the points that needed to be made. The *Toronto Daily Star* opined that "the fire-scourged area will bear the scars of 1911 for a half century to come … The destruction of several mining towns and camps in Northern Ontario by forest fires brings into strong relief the necessity of a more comprehensive and efficient policy of forest conservation."[27] *The Globe* wrote that this "direful conflagration" should "shock the Province into demanding adequate means of protection." The paper did not place the blame on God or on Indigenous people but on prospectors, "who are inclined to regard fire with indifference"; on settlers "who are unpardonably careless in starting fires to clear their land"; and on railroad companies and pulp mills that leave "slash that becomes almost explosive in dry weather."[28] No paper at the time believed that fires "just happen like Spanish influenza."

Still, the *Toronto Daily Star* wrongly saw "a silver lining" in this catastrophe. "Country Is Swept Bare and This Disaster Can't Recur."[29] *The Globe* noted how the folks of Cochrane "believe that there is a good time ahead."[30] Good times were ahead, but not for very long, because politics kept trying to square the fire triangle. There was just too much wealth, not to mention bribery and personal financial interest, for politicians to take on the railroad, mining and timber industries, and the pulp and paper giants that thrived after 1900 because of the demand for newsprint in the United States. Most of these giants were largely owned by American interests.[31]

The Matheson Fire of 1916 should have been Canada's Big Burn, a catastrophe so tragic that governments would have dutifully done what the US was doing to take control of the fire problem. That did not happen. What Canada got was more destructive fires such as those that burned in the prairie provinces in 1919 and the Haileybury Fire which raged through northern Ontario in 1922.

The Globe and other papers had had enough when Haileybury burnt itself out. "We were warned," the paper noted in a 1922 editorial. "The season [1911] had been one of exceptional activity in building, clearing and lumbering and settlers, miners and lumbermen all contributed to the disaster … *The Globe* at that time recommended stringent provisions against all these dangerous practices, and also better fire ranging. How much has been done? How much have we profited from the terrible experience of 1911? The disaster itself shows that a solution has not been found … We question whether the Government has grappled with the problem in a resolute comprehensive way as it would if the country were threatened with invasion."[32]

In the years that followed the Haileybury fire of 1922, the media's crusade did more to further the cause of wildfire management than all the expert reports that the Commission of Conservation had published up until 1921, the year it was shut down. Editors did this in traditional ways but also by adding another feature to their in-depth coverage and editorials. Letters to the editor empowered readers and experts to voice their opinions in ways that would have not otherwise caught the attention of decision-makers. Those that were published by *The Globe* following the Haileybury fire suggested that the government might not have been in synch with what the public was thinking. None of the letter writers believed that Indigenous people were the problem, as the government had so often suggested.

Angus Brabant, fur trade commissioner of the venerable Hudson's Bay Company, wrote in to insist that it wasn't "Indians" who were responsible for forest fires; it was campers, tourists, hunters, and transients.[33] Brabant spoke French, Cree, Ojibway, and Chipewyan. His reputation in Canada was so "enviable" that the company's well-read magazine, *The Beaver*, profiled him in one of their issues. A man by the name of S. Lawrence, who identified himself as an "old-timer," noted that he had "lived through all the big fires of the North and innumerable small ones." He suggested that instead of keeping "Indians" cooped up on reservations, the government should employ them to take care of the forest. Not only would this increase the value of the forest, "it would transform thousands of our Indians into strong, self-reliant men, giving them the conviction that they were guardians of the forests of their fathers, and one of our great natural assets."[34]

Fred Brown, a thirty-year veteran of the forestry industry, acknowledged that railroad companies were the problem, but he also pointed a finger at the government for "not enforcing the rules" that would stop humans from carelessly starting fires."[35] In one of his many letters to *The Globe*, the University of Toronto's dean of Forestry, C.D. Howe, suggested the fire problem was a "question of public morals." He predicted that Ontario would eventually run out of merchantable lumber if nothing was done to stop fires from burning down so much forest. "Over 95 per cent of all fire are due to human carelessness," he said. "We will never have adequate fire protection until we can educate the fool with his match, or until we can educate carelessness out of the human system."[36]

The most prolific letter writer was Frank Barnjum, the pulp and paper magnate who made a name for himself in the press in 1922 by offering two $1,000 prizes for the best essays on practical forestry. To add legitimacy and significance to this competition, he convinced C.D. Howe, Robert Campbell, the director of the Dominion Forest Service, and provincial foresters from New Brunswick and Quebec to be the judges.

In 1923, *The Globe* published approximately seventy-five articles on the subject of wildfire. The stories covered topics such a fire in New Brunswick that threatened to be as destructive as the one in Miramichi in 1825, the decision to extend the fire range patrol season in northern Ontario "until every vestige of fire hazard is passed," and the fate of the Ontario Fire Relief Committee, overdrawn and badly in need of funds to help the destitute. The paper reported on the mining town of Anyox, BC, where hundreds were "blackened and burned" before they got to the safety of a steamship. White River and Wawa came close to burning, *The Globe* reported. Sully, Les Etroits, and Whitworth in Quebec weren't so lucky. All three towns ended up in the line of fire before the winds shifted at the last minute and sent the fire in a different direction.

There was no let-up to this kind of coverage in *The Globe* and other big city papers. In 1924, *The Globe* ran roughly a hundred in-depth articles on wildfire.

26.3 *The Scourge of Canada: Fighting a Forest Fire*, by Cyrus Cuneo, in *The Illustrated London News*, 22 July 1911.

They included stories on fires in British Columbia, Alberta, Quebec, and northern Ontario. Fires, the paper noted, were caused by berry pickers, hunters, lumberman, and by railway companies, which continued to be a big source of trouble. Citing a report from the Board of Railway Commissioners that would otherwise not have seen the light of day, *The Globe* reported that railways were responsible for 63 per cent of the more than $1 million in fire damage in 1923. Three of every four fires reported in 1923 were caused by railways.

Thanks to *The Glob*e and other papers that reported on the British Empire Forestry conference held in Ottawa in 1923, the public learned of Edmund

Zavitz's concern for the practice of mining companies burning bush to expose minerals, and P.Z. Caverhill, chief forester for the province of British Columbia, suggesting a need to get "Canada as a whole," and possibly the entire empire, to solve the fire problem."[37]

Had there been an award for meritorious public service in journalism at that time, as there would be later, *The Globe* would have won it that year. It ran several editorials and opinion pieces on the subject of wildfire, including one that appeared in the "Women's Pages." Writer Edith B. Henderson praised the government of Ontario's decision to finally purchase a fleet of air boats for fire detection. But she blamed lumber companies that disregarded slash laws and "slaughtered forests." "We must keep our trees," she wrote. "They protect us in winter from the chilling blasts and in summer from the burning sun. They are beautiful, useful, generous friends of man. I cannot imagine a land without trees."[38]

It wasn't just women that *The Globe* was aiming for. Editors also aimed their sights on children by sponsoring a "Save the Forest Week" essay-writing contest, challenging students to call attention to the destruction that forest fires caused. The contest drew 329 essays. The winner was Georgina Green from Toronto. Even the Nancy Durham column, which focused on so-called "women's issues," devoted space to the subject.[39]

All this coverage in *The Globe*, the *Toronto Daily Star*, *Edmonton Bulletin*, and many other papers pressured governments throughout Canada to invest more in airplane fire detection patrols, rail velocipedes, power speeders, portable water pumps, fire towers, and men on the ground preventing and supressing fires. Media attention may well have had something to with railway companies, the Board of Railroad Commissioners, and government forest services cooperating in preventing and supressing fires. Between 1922 and 1940, the number of railroad fires dropped steadily, albeit not enough to stop them from being a major source of forest fires in Canada.

Newspaper coverage of wildfire was mostly good news for Ernest Finlayson, who took over as head of the Dominion Forest Service in 1924. Prime Minister Mackenzie King sponsored a National Forest Protection conference that year, undoubtedly aware that there were political points to be earned, given public interest and concern. King hinted that his government might provide the provinces with financial aid in fighting fires. His government also began to provide the funding needed for the Dominion Forest Service to hire more talented professionals.

Had Finlayson been an astute strategist like Pinchot, Henry Graves, William Greeley, Robert Stuart, and other US Forest Service chiefs, he could have used the media to showcase what his agency was up to. But once again, it was not the Canadian way. Finlayson's name appeared in *The Globe* just three times while he headed the Dominion Forest Service between 1924 and 1936.

Most Canadians didn't know he existed. His death by apparent suicide in 1936 didn't even merit a mention in the paper.

Finlayson never really had much of a chance to turn the Dominion Forest Service into a powerhouse with the support of the media, which was sympathetic to better fire management but unclear about how a national strategy might come into play. Mackenzie King, prime minister for most of the 1920s, had little interest in forestry. He was tired of dealing with western provinces that were increasingly resentful of the federal government's management of natural resources. No longer interested in this tug-of-war, King relented to their demands and between 1929 and 1930, quietly negotiated a deal to transfer control over federal forest reserves and other natural resources to Alberta and Saskatchewan. (Manitoba had been given control sixty years earlier.) Earth-shattering as that decision was from a constitutional perspective, it was, as *The Globe* noted in burying the three-paragraph story on page five, "unmarked by ceremony,"[40] as King purposely downplayed the ramifications. Almost overnight, responsibility for preventing and suppressing fires outside of national parks, First Nations, and military reserves, was turned over finally and completely to the provinces. King conceded defeat without much a fight.

The implications played out over the next two decades. As Finlayson pointed out, the provinces were happy to take control over the forests but unwilling to do what was necessary to live with fire. Experts in the field were appalled. "Since Confederation, we have burned five times as many trees in Canada as have been cut by all the lumbermen who ever laid an axe to our forests," Robson Black, president and general manager of the Canadian Forestry Association, told *Maclean's* magazine in 1948, when the Mississagi Fire burned 162,000 hectares of pine, spruce, and balsam, and killed thirteen people. It was a fire so big and so badly out of control that scientist K.E. Pettit of the Dominion Meteorological Services was brought in to experiment with cloud seeding. In these experiments, planes dropped loads of dry ice on a cloud bank. When that didn't work, an unnamed firefighter told *Maclean's* magazine in the same story that "if God ain't too busy tonight he'd better send us rain. Or there ain't going to be no Mississagi timber left. And no camps, no game, no fish. No nothing but the biggest, blackest graveyard you ever seen."

There was, however, a limit to the way newspapers and magazines covered wildfire. Sometime around 1951, when the *New York Times* published the last of a series of articles on how successfully seven east-coast states and two Canadian provinces were working collaboratively to suppress fires, the tone of newspaper coverage on both sides of the border changed.[41] Newspapers continued to report on fire. But because of more sophisticated detection and aggressive suppression strategies, fires were no longer killing so many people, forcing mass evacuations, and destroying towns as they so often had done. Few newspaper reports were longer than four paragraphs. Without actually stating it, newspapers tacitly acknowledged that the war on fire had been won. The enemy was still there, but they believed that there were enough firefighters on the ground, in lookout towers, and in planes detecting fires to keep the public from harm.

In the next thirty-seven years, there were just four wildfires, all of them in California – the Rattlesnake Fire of 1953 in the Mendocino National Forest, the Inaja Fire of 1956 in the Cleveland National Forest, the Coyote Fire of 1964 near Santa Barbara, and the Laguna Fire of 1970 in San Diego County – that reminded Americans that wildfire was still a deadly force. The Chinchaga Fire of 1950 was also a head spinner, but it did not kill anyone. Not until the Yellowstone fires of 1988 did it become clear, as journalist Harrison E. Salisbury pointed out in the *New York Times*, "that man does not possess the power to control natural forces so vast as these fires."[42]

It was difficult, however, for the media, the public, and decision-makers to embrace the idea that prescribed burns and allowing lightning-triggered fires to burn themselves out in remote areas was the way to move forward, as many wildfire and forest scientists started to think. Undoing three quarters of a century's thinking that detection and suppression was the answer was difficult, if not impossible. When Yellowstone burned, President Ronald Reagan dismissed the relatively new "let it burn" strategy in the park as "cockamamie." Interior Secretary Donald Hodel described it as "absurd," and ordered national park officials to fight all fires from that point on.

The Yellowstone fires made it onto the front page of every major newspaper in the United States. Everyone thought a national icon was destroyed, never to come back again. Papers such as the *Los Angeles Times* and *New York Times* tried to make ecological sense of the need to "let it burn," or "leave to nature" wildfire management strategy. But public support for the strategy of suppression prevailed almost everywhere, partly because of past practices, but also because some prescribed burns, such as the Cerro Grande Fire of 2000, got out of control, and because national park officials on both sides of the border feared that smoke from these fires would hurt tourism.

It was only a matter of time before a new recipe for disaster – climate change, decades of successful fire suppression, and too many people moving into and working in forested ecosystems – overtook the ability of firefighters to control wildfire. It's fair to suggest that much of the media has been slow in connecting the present with the past in a meaningful way. The fact that the role of climate change was still being debated in 2021 – thirty years after Mike Flannigan and Charlie Van Wagner published a paper predicting that fires would burn more often in a rapidly warming world – is a testament to that. But it is more complex and nuanced than that.

The *Globe and Mail*, for example, a right-of-centre paper, published more than 125 articles on the 2016 Horse River Fire that forced the evacuation of more than 88,000 people from Fort McMurray. But only three of those articles discussed the possibility that climate change might have played a role – inexcusable, considering that as far back as 1989, Canadian wildfire scientist Charlie Van Wagner suggested to the same paper that the fingerprints of climate change were all over the record-breaking fire season in Manitoba that year.

But who could blame the editors, when Prime Minister Justin Trudeau downplayed the role of climate in the Fort McMurray fire, as did NDP leader Jack Layton, who insisted that this was not the time to talk about the subject or to lay blame? Even Green Party leader Elizabeth May felt compelled to clarify a remark she made to reporters that "of course" the fire was linked to climate change.[43] Shortly afterwards, May noted in a press release that "some reports have suggested that wildfires are directly caused by climate change. No credible climate scientist would say that and neither do I."

As far back as 2015, a group of scientists reported what was obvious to almost everyone in the wildfire science community: that fire seasons worldwide were lengthening and that "fire weather" – the combination of humidity, temperature, wind, and other factors that help blazes to spread – was becoming more extreme. What was missing, according to those scientists, was the global mapping and monitoring necessary to make definitive statements about trends.[44] So instead of saying that fires like the one in Fort McMurray were linked to climate change, scientists began framing intense fires as ones that fit a trend in a rapidly warming world.

This was technically correct, but ask anyone who lost their home to wildfire whether climate change is real and whether it accounts in some way for intense fires and the answer is likely to be the same as that given by John Haugen, deputy chief of the Lytton First Nation. "It's [climate change] real in my books," he told the media, months after he had a chance to start putting his life back together.[45]

This tiptoeing around the subject of climate change by politicians and even some scientists would never have happened in the days when Gifford Pinchot was leading the charge. Pinchot pointed out that nothing permanent can be accomplished unless it is backed by sound public sentiment. Arousing a general interest in and gradually changing public awareness about how to live with fire will only occur when the media is on the same page with scientists. Suppressing fire, which the media has tended to focus on for so long, only encourages people to believe that the "business as usual" approach will be successful. Doubting the connection between climate change and wildfire, as many conservative media outlets continue to do, only makes them resist new approaches. The odds against societies winning the battle against large and increasingly intense wildfires continue to be stacked against them. Even though progressive newspapers may extend their reach to the public and sound the alarm the way they did in the past, laissez-faire governments continue to be willfully slow to step in, content in the belief that free markets and fire suppression can solve any unforeseen problems that arise.

CONCLUSION

On 30 June 2021, a wildfire ignited south of the British Columbia town of Lytton, killing two people, injuring several others, burning down most of the buildings, and forcing the evacuation of more than 1,500 people in and around the surrounding area. When the smoke began wafting in, according to Mayor Jan Polderman, it took all of fifteen minutes before the town was ablaze.[1] But no one saw the fire coming.

This was the third time in five years during Premier John Horgan's tenure that catastrophic wildfires had taken their toll. "I cannot stress enough how extreme the fire risk is at this time in every part of British Columbia," he told members of the media the day after the evacuation. "This is not how we usually roll in a temperate rainforest."[2]

Once again, a political leader was out of synch with what was happening in the forests. Lytton is actually located in the drier, fire-prone montane region – which dominates much of the interior of the province. Montane forests are those that grow in high elevations. Contrary to what Horgan said that day, this is exactly how things have been rolling since at least 2003, when more than 45,000 people were evacuated from the Kelowna and Kamloops region area of the province as fires tore through thick stands of cedar-hemlock, Ponderosa pine, Douglas fir, and Engelmann spruce.

The year 2003 was notable not for the amount of forest that was consumed in Canada and the United States but for the number of people who were in harm's way. Twenty-five fires that burned 13 per cent of Glacier National Park along

the Alberta-Montana border that summer sent bus and car tourists packing so quickly that those who were eating at a historic lodge didn't have time to finish their meals. Two thousand people in Crowsnest Pass along the Alberta-BC border spent a good part of their days watching the fires from a place that was not their home. Jasper National Park burned big and scary enough that officials feared fire might spread to the towns of Jasper and Hinton. California's Cedar Fire was the most destructive fire in the state's history up until then. Banff would have burned bigger and possibly in catastrophic fashion had wildfire managers fighting a fire in neighbouring Kootenay National Park not thrown away the playbook and lit a backfire that stopped it before it jumped into Banff. Parks Canada firefighters named the blaze "The Holy Shit Fire," because that's what a slew of senior government bureaucrats exclaimed when they were flown in to have a look at it.

C.1 As fires burn bigger and hotter, more towns are being destroyed. The Kincade Fire in Sonoma County, 2019. Photo by Kari Greer.

Parks Canada's fire specialist Rob Walker was prescient in suggesting at the time that this fire, coming as it did after the record-breaking Yellowstone and Manitoba fires of 1988 and 1989, the deadly 1994 fire season in the United States, the nuclear catastrophe that almost came with the Cerro Grande Fire of 2000, the pyroCbs-inducing Chisholm Fire of 2001 in Alberta, and the calamitous Hayman Fire of 2002 in Colorado, was a harbinger of future events. The following summer, the Yukon and Alaska burned like they had never burned before. The 2007 Murphy Complex Fire in Idaho was the state's largest since the Big Burn of 1910. In 2007, the tundra on Alaska's North Slope burned like it hadn't for at least 10,000 years. Fires in West Kelowna in 2009 caused more damage than they did to the region in 2003. In 2011, the entire town of Slave Lake in Alberta was forced to evacuate as fire swept in and destroyed a third of the structures. That same summer, the Richardson Fire came almost as close to wiping out Fort McMurray as the Horse River Fire did in 2016, when it forced the evacuation of 88,000 people. In the next five years, these shocking events were overshadowed by as big or bigger blazes in Siberia, Australia, California, and much of the western United States. By the end of the 2021 fire season, the 2003 Cedar Fire was no longer the most destructive fire in the state. It was by then only the ninth.

Many things have been in play that have made the situation much more volatile since 2003. Successfully fighting fires for as long as Canada and the United States have, sometimes in tandem, has resulted in aging forests that are not as a resilient as are young mixed stand of trees and shrubs. Peatlands – bogs, fens, swamps, and marshes that can stop or slow a wildfire – have been systematically drained or degraded to make way for farms, roads, hydro dams, energy developments, small towns, and acreages. When degraded peat ignites, it is almost impossible to extinguish without the help of heavy rain or snow.

In addition, low property values and the allure of rural life have continued to entice retirees and other people from big cities to move into forested environments. Many newcomers have built homes out of cedar and surrounded them with wooden fences and highly flammable ornamental plants. To suppress the growth of unwanted vegetation, they have laid bark on top of their gardens. Some spend their workdays or weekends in the forest on all terrain vehicles that have exhaust systems that can get hot enough to start a fire. Trains they hear at night occasionally shoot sparks into the dry vegetation along the tracks, just as they did in the past.

The warming climate that American scientist Jim Hansen warned the world about in 1981 was also impactful. It was on full display between 2011 and 2021, the hottest decade on record. In many places, the fire season during that period was extended by several weeks. Lightning strikes continued to be on the rise. Thick smoke rising from intense fires appeared to be arresting the ability of

C.2 The Telegraph Fire, Arizona, 2021. Photo by Andrew Avitt.

clouds to produce rain. The total number of fires wasn't the problem; it actually decreased in most places. It was the megafires that put so many people in harm's way.

Each year brings new surprises such as giant heat domes, fire tornadoes, and multiple pyroCbs that have enough energy, as they did in British Columbia in 2021, to trigger 72,000 pulses of lighting. Wind-driven fires like the one that raced across Waterton Lakes National Park in less than eight hours in 2017 are moving with speeds once relatively rare. Greenland wasn't supposed to burn until it did, in the summers of 2017 and again in 2019. Peat fires like those that smouldered in Yakutia in the summer of 2021 sent smoke 3,000 kilometres to the North Pole for the first time in recorded history. In a six-week period in 2021, the fires in Yakutia released about 150 megatonnes of carbon dioxide – equivalent to the amount that fossil fuels in a country like Venezuela emit in an entire year.

In the past, firefighting agencies in Canada and the United States could count on each other to share resources, as the Fire Compacts allowed them to. If those resources were tapped out, they could call on firefighters from Australia, South Africa, and other countries, who had little to do during their winters. But so many countries – Canada, the United States, Russia, France, Greece, Italy, Turkey, Israel, Lebanon, and others – burned so ferociously in 2021 that there was not a lot to share. The COVID pandemic made things even more complicated. Finally, the fire seasons of California and the west were beginning to overlap with those in Australia.

There is the possibility of a fire season in the future that spares none of the fire-prone provinces or states. That has never happened before.

Big Canadian cities such as Vancouver, Victoria, Montreal, and Toronto are not as likely to burn as the river valleys in the hearts of Edmonton, Saskatoon, and other cities might if fire is ignited in the middle of the night during a long hot drought such as the one that parched the west in 2021. But smoke could make life in such cities miserable, as it did in 2021 when a pall from the California, Oregon, and British Columbia fires drifted east to Toronto and New York City. Soot could seriously compromise water quality for tens of millions of people, as it did in Canberra in 2003, in Fort McMurray in 2016, and in Colorado in 2002.

The most vulnerable people are those who live and work in forested environments where the avenues of evacuation are few and where there is no place to take refuge from smoke. Even if they wanted to, most rural communities, especially those inhabited by Indigenous people, do not have the

C.3 Firefighting resources are increasingly in short supply as fires spread around the world. Photo by Kari Greer.

C.4 The Kenow Fire, 2017. Parks Canada used all the resources it had to save the town of Waterton Lakes and the historic Prince of Wales Hotel. Parks Canada photo.

tax base to manage the vegetation in and around town the way that more affluent communities such as Waterton, Banff, and Jasper do because they receive the federal funding that comes with being located in a national park. This is one of the reasons why First Nations communities such as Lytton account for 40 per cent of the evacuations in Canada. Their welfare is treated as less important than that of tourists who spend big money in national parks. But as the Kenow Fire of 2017 and California's Caldor Fire of 2021 demonstrated, even national parks like Waterton and resort towns like Lake Tahoe are not necessarily safe.

This is not the "new normal," as meteorologist and wildfire scientist Mike Flannigan likes to say in correcting the media. There is nothing "normal" about what is happening, and nothing much that firefighters can do, in many cases, with the traditional tools and strategies they have relied on for so long. Prior to the 2021 fire, Lytton was baking in temperatures close to 50 degrees Celsius for three days straight. That's five degrees shy of the world record set in the Furnace Creek area of Death Valley the year before. The big difference between Death Valley and a rural town like Lytton is that Death Valley has few trees to speak of; 12,000 years of gradual warming has transformed it from a wetland fed by melting glaciers to a desert where most of the springs and ephemeral streams are sustained by 8,000- to 12,000-year-old fossil water still trickling underground from those long-gone glaciers. Some relics of the past, such as pupfish, have hung on, but just barely.

Fifty degrees, forty degrees, and even temperatures in the low to mid-thirties are all that a wildfire needs to behave dangerously and often unpredictably, especially when driven by strong winds, low humidity, and plenty of parched trees and dry grass on the ground. Many such fires have become meteorological events that cannot be stopped. On a hot windy day, once a fire reaches the size of a football field, as Flannigan likes to say, "the horse is out of the barn." Fire retardants and water bombers may be able to slow a fire or steer it away, but they can't extinguish it when it burns big and is driven by high winds and low humidity.

The impact of wildfire is no longer just regional, national, or even continental. It has become a global phenomenon with the potential to rival the impact of the Little Ice Age, which was global in scale and freakish in the way it drove hurricanes, intensified drought, and iced up temperate maritime climates such as the one which keeps most of England cool, moist, and free of snow. "Nature's Mutiny" is how historian Philipp Blom describes it in his book with that title. The Little Ice Age was the opposite of our current climate crisis in that it cooled rather than heated the earth, but it was eerily similar in impact to what is happening now because it was the one time in 5,000 years of recorded history when humans got a sense of what happens when the climate is no longer stable. The cooling began in 1302, triggering a five-year drought and weather patterns like those in 2018 when continental Europe suffered through exceptional heat

and withering drought, and in 2021, when the heat dome over much of western North America would not budge.

C.5 The destruction of power lines has resulted in brownouts and blackouts in many parts of the west. Photo by Kari Greer.

Transitional phases in climate are often characterized by periods of low variability in which unwelcome weather patterns remain stable for a long time. The Great Famine of 1315 to 1321, considered to be the largest European-wide food crisis of the past millennium, was an example of how this played out. Severe winters and wet, cool summers resulted in widespread crop failures and difficulty in curing straw and hay for livestock, as well as obtaining salt to preserve meat because brine does not evaporate well in cold, wet weather. Food prices doubled in some places. Even Edward II, the king of England, had trouble finding bread for his entourage while travelling through St Albans in the spring of 1315. Hunger was so rampant in some parts of Europe that people had to guard graves from being pillaged. Many densely constructed cities like Siena burned so often and severely that laws were put into place to oblige citizens to place buckets of water outside their front doors.

We are not living in medieval times. Most people back then saw fire as an act of God or the product of spontaneous combustion. Nor are we as clueless about wildfire as most of those who experienced the Dark Day at Noon in 1780, the Miramichi Fire in 1825, the Ottawa Valley and Saguenay River Fires of 1870, and the Peshtigo Fire that burned so big the following the year. But continuing to assume that we can suppress fire as long as there are enough resources to put one out is akin to playing Russian roulette. At some point, the reckless gambling that has been going on is going to result in more of the destruction that we saw in Fort McMurray in 2016, Australia in 2019–20, British Columbia in 2021, and in California, Oregon, Washington, and Colorado for much of the past decade.

Fires are now so propulsive that they have brought communities to their knees, closed highways and flyways, and forced major utilities into bankruptcy. They have turned the lazy, hazy days of summer on the beach, by a lake, or in the mountains into noxious, smoke-filled ones that are best spent inside with air purifiers. In just fifteen months in 2020–21, fire killed or mortally wounded between 13 and 19 per cent of the world's giant sequoias, trees that can grow over hundreds of years to more than 80 metres in height and eleven metres in diameter.

The reluctance to invest more in fire science and in making small towns, cities, and Indigenous communities more resilient to fire with tax incentives, tougher building codes, and the resources needed to manage the landscape as Indigenous people once did is unconscionable, considering how much is being spent on suppression and on compensating victims of fire. In 2017 and 2021, British Columbia spent more than $500 million on suppression – twice the ten-year average. The total cost of the Horse River Fire was close to $9 billion. Despite these expenses, in 2019 the Alberta government eliminated the fire science coordinator position, reduced staffing for detection towers, eliminated the Wildland Firefighter Rappel Program, and rid itself of the cost of one air tanker unit.

In the 1980s, the Canadian Forest Service employed more than 2,000 people. In 2021, there were fewer than 800. A small fraction were wildfire specialists. In 2020, the Canadian government announced a multi-billion-dollar plan to plant two billion trees over the next decade, when it would have been wiser and more cost-efficient to also allocate the funds for restoration of peatland: those bogs, fens, swamps, and marshes that can stop a wildfire in its tracks and store much more carbon than any forest. In 2019, the Natural Sciences and Engineering Council began investing in wildfire science, but not nearly enough to fill in the gaps when scientists like Flannigan are lured to one province from another.

The public has been culturally conditioned by historians, writers, artists, novelists, and poets either to ignore fires or to vaguely remember the damage they have done and doggedly assume that investing in fire suppression is a way of taming these dragons in the forest. Collectively, Canadians still have not come to grips with the fact that most of the country's landscape is unforgivingly boreal – fiery, spongy, buggy, running with wolves and bears, swimming with beavers, and snapping with turtles and rattlers denning in peat. Wildfire does not disapprove or approve even in the more temperate regions or the coastal wetlands around the Great Lakes. It has nothing to say except for the "fact of itself," as Margaret Atwood wrote in her novel *Surfacing*. It is merely a rapid-fire chemical reaction that responds to the combustible material around it.

It is not too early to suggest that the warming that is taking place now could have the same impact as what happened at the beginning of the Little Ice Age. When a team of scientists began analyzing events that may have had an impact on the world's climate in 2020, they made sure to consider the pandemic-related lockdowns that dramatically reduced carbon emissions. What they found was surprising, yet it seemed obvious in hindsight. Even though the reduction of emissions during COVID-19 resulted in noticeably clearer skies throughout much of the world, another reason the climate slightly cooled for a time was the runaway fires in Australia.[3]

Those bushfires were the worst in Australia's history, and proving once again that smoke which goes up must come down, but not necessarily in the region from which it originated. Like the multiple pyroCbs in British Columbia and Washington State in 2018 that sent smoke as far away as Greenland, the dark plumes from the Australian bushfires rose above the troposphere, where weather is created, into the stratosphere. For a time, this resulted in dark days in Chile and red skies in Buenos Aires. Some of that smoke, according to the United Nations Meteorological Association, likely drifted into Antarctica. As those aerosols circled the southern hemisphere, sulphates and other particles interacted with clouds, making their droplets smaller and better able to reflect solar radiation back into space.

We are not going to stop fire, just as we cannot stop hurricanes and atmospheric rivers from dropping rain down upon us. Big destructive fires will continue to punish us even if we slow fossil fuel consumption, as we need to do. The climate will continue to heat up in ways that will double the amount of lightning in many places.[4]

Understanding the past is essential if we are to anticipate the future. Learning lessons from past fires, however, will only go so far in preparing us for what's to come because the historical record has always been incomplete. Moreover, the past is no longer a reliable proxy for what is ahead because the ingredients that create a catastrophic fire now are much different than they have ever been before and the landscape itself has been altered because of climate change and human development.

Wildfires will continue to burn bigger and hotter because the jet stream isn't what it used to be. This fast-flowing river of air snakes around the Northern Hemisphere at altitudes of up to six kilometres, creating high and low-pressure areas and dragging weather and smoke along with it. It is powered by the difference in temperature between cold Arctic air and the warmer air to the south. As the Arctic warms and the temperature difference diminishes, the stream's westerly winds weaken, vying with La Niña and El Niño as a force of nature that shapes weather patterns from Alaska down to California.

C.6 Prescribed burning and thinning the forest close to towns and assets are necessary to make towns and assets resilient to future fire. Photo by Kari Greer.

Just such a shift was associated with the "ridiculously resilient ridge"[5] that hunkered down over most of California in the fall of 2017, persistently redirecting wet weather northward to Alaska, which at one point got a record 100 millimetres of very wet snow and rain in twelve hours. This was why California firefighters were fighting fires over the Christmas holidays; the fall rains didn't come.

What's needed is the political will, energy, and ingenuity that made Gifford Pinchot so successful at wresting control of forests away from industry and using science and technology to balance profit and lifestyle with conservation imperatives. Letting fires burn, as the US National Parks Service did in Yellowstone in 1988, is one solution. But the forests now are so littered with towns, acreages, trailer parks, oil and gas rigs, power lines, and roads that this is difficult and dangerous to do.

Prescribed burning, such as that which was once practiced by Indigenous people, is more promising because such fires can be controlled. Thinning the forest around communities can slow blazes down long enough for firefighters to

move in. Better building codes can make forested communities more resilient. And power companies have ways in which to prevent their lines from sparking and/or falling down in high winds. Decarbonizing will help those who have to deal with fire in the future. All of these approaches will help, but the fact is that there is no single strategy that will help us to live comfortably with fire.

Pinchot and his colleagues were wrong to demonize fire as aggressively as they did for nearly a century. Their pyrophobia set up the forests and grasslands in North America to burn as big and intensely as they have been doing. But they at least recognized that the "business as usual" approach to managing the forests of North America needed to change, and that science was needed to come up with the answers.

If politics continues to square the fire triangle, as it has so often done, good fires will continue to be overtaken by destructive ones and it will be more difficult, if not impossible, to live with fire. The Dark Days that so terrified people in 1780

C.7 Forest firefighters in the Los Padres National Forest in California in 2008 trying to save an oak rumoured to be more than a thousand years old. Photo by Kari Greer.

will become common and bring more misery. We all won't "go down," as Adrien Welsh feared while taking flight from Fort McMurray with her children in 2016. Some people survived the sinking of the *Titanic* and lived to tell about it, just as all but two people living in Fort McMurray did, and as most people in British Columbia, California, Colorado, and other regions have since the turn of the century when megafires became catastrophic. But there will be *Titanic* events and many more "dark days at noon" if we continue to binge on fossil fuels, degrade the landscape in ways that favour fire, and construct buildings next to mature forests that will inevitably burn on hot windy days when lightning or an arsonist strikes, a campfire is left unattended, or equipment malfunctions.

The Horse River Fire of 2016 demonstrated this, as McMaster University scientist Mike Waddington and PhD candidate Sophie Wilkinson discovered. It was a hot, dry, and windy day when an all-terrain vehicle likely sent sparks into the forest outside of town. Local buildings were not designed to be resilient to the flying embers that blew in from two kilometres away and pyroCbs bedevilled wildfire fighters. The most surprising revelation was the pace and intensity of the fire as it passed through. This wetland had been drained in an experiment to grow trees. Had it still been water-soaked, Waddington and Wilkinson suggested, it might have slowed the fire long enough for firefighters to prevail. Instead, the slow-burning peat magnified and prolonged the fire.

It's not like we haven't been warned. Fort McMurray nearly burned to the ground in 2011 when the Richardson fire torched more than 600,000 hectares in northern Alberta. The California town of Paradise was threatened thirteen times before it was levelled in 2018.

More "Dark Days" are coming to this rapidly warming world unless we develop a culture that respects the element of fire as Indigenous people did. Fire has nothing to say except for the fact of itself. It will always be quite simply a chemical reaction, a propulsive oxidation of hydrocarbons shaped by terrain, weather, climate, and the combustible material around it. We have to learn to live with fire and find ways of containing it. Fire will not learn to live with us.

Illustrations

Notes

Introduction

1 Donald Creighton, *Canada's First Century, 1867–1967* (Toronto: MacMillan, 1970).
2 Erin L. Moss, Henderikus J. Stam, and Diane Kattevilder, "From Suffrage to Sterilization, Eugenics and the Women's Movement in 20th Century Alberta," *Canadian Psychology/ Psychologie canadienne* 54, no. 2 (2013): 105–14.
3 Donald A. Wright and Donald Creighton, *A Life History* (Toronto: University of Toronto Press, 1915).
4 B. Alexander and Michael Keating, "Colorful Historian Did Pioneering Work," Obituary, *Globe and Mail*, 8 January 1988, 9.
5 D.G. Creighton, "Review of *The North American Assault on the Canadian Forest: A History of the Lumber Trade between Canada and the United States*," by W.A. Carrothers, in *The Canadian Journal of Economics and Political Science* 5, no. 2 (1939): 243–5.
6 Mary R. Gunstra and David L. Martell, "A History of Railway Fires in Ontario," *Forestry Chronicle* 90, no. 3 (2014): 314–20.
7 Alan MacEachern, *The Miramichi Fire: A History* (Montreal and Kingston: McGill-Queen's University Press, 2020), 141.
8 A.R.M. Lower, *The Northern American Assault on the Canadian Forest* (New York: Greenwood Press Books, 1938), 76.
9 W.F. Rannie, "The 'Grass Fire Era' on the Southeastern Canadian Prairies," *Prairie Perspectives: Geographical Essays* 4 (2001): 1–19.
10 Peter C. Newman, *Caesars of the Wilderness* (Markham, ON: Penguin Books Canada, 1987), 53.
11 I.S. MacLaren, "Notes Towards a Reconsideration of Kane's Art and Prose," *Canadian Literature* 113–114 (Summer–Autumn 1987): 179–205.
12 Susanna Moodie, "Burning the Fallow," *Roughing it In the Bush*. Project Gutenberg EBook. https://www.gutenberg.org/files/4389/4389-h/4389-h.htm#link2HCH0016.
13 Susan Glickman, *The Picturesque and the Sublime: A Poetics of the Canadian Landscape* (Montreal and Kingston: McGill-Queen's University Press, 1998), 47.
14 F.R. Scott, "New Poems for Old: The Revival of Poetry," *The Canadian Forum* 11 (June 1931): 337–9.
15 Stephen J. Pyne, *Awful Splendour: A Fire History of Canada* (Vancouver: University of British Columbia Press, 2007), 478.
16 Ibid., xxvi.
17 Robert Bott, Peter Murphy, and Robert Udell, *Learning from the Forest: A Fifty-Year History in Sustainable Forest Management* (Calgary, AB: Fifth House, 2003), 44.
18 Peter J. Murphy and Robert E. Stevenson, "A Fortuitous International Meeting of Two Yale Foresters in 1908," *The Forest Chronicle* 3 (May–June 2014), 321–3.
19 John Wesley Powell, *Report on the Arid Regions of the United States* (Washington, DC: Government Printing Office, 1878), 195.
20 "The Whole of Canada in One Mass of Flame," *Perth Courier*, 26 August 1870, front page.
21 Timothy Egan, *The Big Burn: Teddy Roosevelt and the Fire That Saved America* (Boston-New York: Houghton Mifflin Harcourt, 2009).

22 Donald Worster, "Transformations of the Earth: Toward an Agroecological Perspective in History," *Journal of American History,* 76 (March 1990) 1087–1106.

23 Mike Flannigan and Charlie Van Wagner, "Climate Change and Wildfire in Canada," *Canadian Journal of Forest Research,* 21: 66–72.

Chapter One

1 George Washington, *The Diaries of George Washington, Volume 3*, ed. Donald Jackson and Dorothy Twohig (Charlottesville: University Press of Virginia, 1978), 353.

2 Abigail Adams, "Abigail Adams to James Lovell, 24 May 1780," Adams Papers Digital Edition, Adams Family Correspondence 3, Massachusetts Historical Society.

3 Samuel Phillips Savage Diaries, 1770–1795, Massachusetts Historical Society. https://www.masshist.org/collection-guides/view/fa0532.

4 "The Darkest Day," *The Williams Quarterly* IV (1857): 338–40.

5 Ibid.

6 Thomas J. Campanella, "'Mark Well the Gloom': Shedding Light on the Great Dark Day of 1780," *Environmental History* 12 (January 2007): 37.

7 *Connecticut Historical Collections* 2nd ed (1836), compiled by John Warner Barber, 403.

8 *The Williams Quarterly*, Volume IV, 1857, 337.

9 D.M. Ludlum, "New England's Dark Day: 19 May 1780," *Weatherwise* 25 (1972): 112–19.

10 Richard M. Devens, *Our First Century* (Springfield, MA: C.A. Nichols & Co, 1876), 91.

11 Ezra Stiles, *The Literary Diary of Ezra Stiles,* ed. F.B. Dexter (New York: Charles Scribner's Sons, 1901).

12 Erin R. McMurry, Michael C. Stambaugh, Richard P. Guyette, and Daniel C. Dey, "Fire Scars Reveal Source of New England's 1780 Dark Day," *International Journal of Wildland Fire* 16 (2007): 266–70, https://doi.org/10.1071/.

13 Nerilie J. Abram, Helen V. McGregor, Jessica Tierney, et al., "Early Onset of Industrial-era Warming across the Oceans and Continents," *Nature* 536 (2016): 411–18, https://doi.org/10.1038/nature19082.

Chapter Two

1 Meteorites rarely cause forest fires, as some wrongly assumed when one that fell from the sky near Woodstock, New Hampshire, in 2017 coincided with a small forest fire. But the massive one that crashed into Mexico's Yucatán Peninsula sixty-five million years ago could have transferred its kinetic energy to thermal energy. See Adam Mann, "Life After the Asteroid Apocalypse," *Proceedings of the National Academy of Sciences of the United States of America* 115, no.23 (5 June 2018): 5820–3.

2 Erin R. McMurry, Michael C. Stambaugh, Richard P. Guyette, and Daniel C. Dey, "Fire Scars Reveal Source of New England's 1780 Dark Day," *International Journal of Wildland Fire* 16 (2007): 266–70, https://doi.org/10.1071/.

3 Miron L. Heinselman, "Fire in the Virgin Forests of the Boundary Waters Canoe Area, Minnesota," *Quaternary Research* 3 (1973): 329–82.

4 David H. Jurney, "Anthropology of Fire in the Ozark Highland Region," in *Proceedings of the 4th Fire in Eastern Oak Forests Conference*, 17–19 May 2011, ed. Daniel C. Dey et al., (Newtown Square, PA: USDA Forest Service, 2012), 12–33.

5 O. Pechony and D.T. Shindell, "Driving Forces of Global Wildfires over the Past Millennium and the Forthcoming Century," *Proceedings of the National Academy of Sciences* 107, no. 45 (November 2010): 19167–70.

6 Andrew Barton et al., "Reconstructing the Past: Maine Forests Then and Now," *Northern Woodlands Magazine* (Summer 2013): 40–8.

7 Christopher I. Roos, María Nieves Zedeño, Kacy L. Hollenback, and Mary M.H. Erlick, "Indigenous Impacts on North American Great Plains Fire Regimes of the Past Millennium," *Proceedings of the National Academy of Sciences* 115, no. 32 (Aug 2018): 8143–8.

8 William E. McClain et al., "Patterns of Anthropogenic Fire within the Midwestern Tallgrass Prairie 1673–1905: Evidence from Written Accounts," *BioOne, Natural Areas Journal* 1, no.4 (October 2021), 283–300.

9 Henry David Thoreau, *The Maine Woods* (Princeton, NJ: Princeton University Press, 1983), 70.

10 Henry T. Lewis, *A Time for Burning* (Edmonton: University of Alberta Press, 1982).

11 William Wallace Scott, *A History of Orange County, Virginia: From Its Formation in 1734 to the End of Reconstruction in 1870* (Orange County, VA: E. Waddey, 1907).

12 Antone Pierucci, "The Ancient Ecology of Fire Archaeology," *Archaeological Institute of North America* (September–October 2017). https://www.archaeology.org/issues/272-1709/letter-from/5826-letter-from-california-fires.

13 Ibid.

14 Stephen Barrett and Stephen Arno, "Indian Fires in the Northern Rockies: Ethnohistory and Ecology," in *Indians, Fire, and Land in the Pacific Northwest*, ed. Robert Boyd (Corvallis, OR: Oregon State Press, 1999), 50–64.

15 Lewis and Clark, *Journals of the Lewis and Clark Expedition* (J20 July 1805), https://lewisandclarkjournals.unl.edu/item/lc.jrn.1805-07-20?.

16 Ibid.

17 David Thompson, *David Thompson's Narrative of His Explorations in Western America, 1784–1812,* ed. J.B. Tyrell (Toronto: Publications of the Champlain Society, 1916), 24.

18 Ibid.

19 Alexander Henry, *The Journal of Alexander Henry the Younger, 1799–1814,* ed. Barry M. Gough (Toronto: University of Toronto Press, 1988), 100.

20 Irene M. Spry, ed., *The Papers of The Palliser Expedition, 1857–1860* (Toronto: The Champlain Society, 1968).

21 W.F. Rannie, "The 'Grass Fire Era' on the Southeastern Canadian Prairies," *Prairie Perspectives: Geographical Essays* 4 (2001): 12.

22 James Raffan, *Emperor of the North: Sir George Simpson and the Remarkable Story of the Hudson's Bay Company* (Toronto: Harper Collins, 2008), 12.

23 Ibid., 13.

24 Ibid.

25 Ibid.

26 Mario Giguère and Bronwyn Evans, eds., *Mon Journal: The Journal and Memoir of Father Léon Doucet OMI, 1868–1890* (Calgary: Historical Society of Alberta, 2018), 113.

27 "Carrying Fire the Pikunni Way," US Forest Service video (30 March 2017), https://www.youtube.com/watch?v=VdLfdjX6smU.

28 Henry F. Dobyns, "Estimating Aboriginal American Population," *Current Anthropology* 7 (1966): 395–449.

29 Nancy Shoemaker, *American Indian Population Recovery in the Twentieth Century* (Albuquerque: University of New Mexico Press, 1999), 22.

Chapter Three

1 Henry David Thoreau, *Walden; or Life in the Woods* (Boston, MA: Ticknor and Fields, 1854).

2 Henry David Thoreau, *The Writings of Henry David Thoreau,* Vol. 5 (Boston, MA: Houghton Mifflin, 1906), 36.

3 Ibid., 21–5.

4 Ibid.

5 Ibid.

6 Loren Eiseley, *Darwin's Century: Evolution and the Men Who Discovered It* (Garden City, NY: Anchor Books, 1961), 353.

7 Henry French, *Farm Drainage: The Principles, Processes, and Effects of Draining Land with Stones, Wood, Flows, and Open Ditches, and Especially with Tiles* (New York: Orange Judd and Co., 1859), 222.

8 Alan MacEachern, *The Miramichi Fire: A History* (Montreal and Kingston: McGill-Queen's University Press, 2020), 90.

9 Ibid., 89

10 Ibid.

11 Anon., *A Narrative of the Late Fires of Miramichi, New Brunswick* (Halifax, NS: 1825) https://archive.org/details/cihm_37196.

12 Ibid.

13 Ibid.

14 Robert Cooney, *The Autobiography of a Wesleyan Methodist Missionary (formerly a Roman Catholic)* (Montreal, 1856), 54–5.

15 Ibid., 56.

16 Ibid.

17 MacEachern, *The Miramichi Fire,* 153.

18 Anon., *A Narrative of the Late Fires of Miramichi, New Brunswick*, 7.

19 Ibid.

20 MacEachern, *The Miramichi Fire*, 91.
21 Catherine Parr Traill, *The Backwoods of Canada* (Montreal and Kingston: McGill-Queen's University Press, 1997), 144.
22 "The Whole of Canada in One Mass of Flame," *Perth Courier*, 26 August 1870.
23 "The Fires in the Ottawa District," *The Globe*, 27 May 1871, 3.
24 Ibid.
25 Ibid., 2.
26 "Fires Ravaged Michigan's Thumb in 1871, 1881," *Michigan News, University of Michigan*, 8 January 2007, https://news.umich.edu/fires-ravaged-michigans-thumb-in-1871-1881.
27 Reverend Peter Pernin, "The Great Peshtigo Fire, An Eyewitness Account," *Wisconsin Magazine of History* 54, no. 4 (1971): 3.
28 Ibid., 3.
29 Ibid.
30 W. Beck, "Substantial and Vivid Novel of Wisconsin," *Chicago Daily Tribune*, 29 December 1957, https://www-proquest-com.proxy.queensu.ca/historical-newspapers/substantial-vivid-novel-wisconsin/docview/180229045/se-2?accountid=6180.
31 MacEachern, *The Miramichi Fire*, 3–9.

Chapter Four

1 J.W. Powell, *Report on the Lands of the Arid Region of the United States*, 2nd ed. (Washington, DC: Government Printing Office, 1879), 17.
2 "Landscape History of Central New England," *Harvard Forest*, 2021, https://harvardforest.fas.harvard.edu/diorama-series/landscape-history-central-new-england.
3 F.B. Hough, *The Elements of Forestry* (Cincinnati, OH: Robert Clark & Co., 1883), 313.
4 "Forest Fires: A Thousand Square Miles, A Blackened Waste," *The Globe*, 10 September 1881, 2.
5 Ibid.
6 "A Saffron-Colored Sky, A Curious Phenomenon and One Widely Extended," *New York Times*, 9 September 1881, 3.
7 "Fatal Forest Fires Rage … People Take to the Rivers to Save Their Lives," *Adams County News*, 17 September 1902.
8 W.F. Rannie, "The 'Grass Fire Era' on the Southeastern Canadian Prairies," *Prairie Perspectives: Geographical Essays* 4 (2001): 1–19.
9 S. Raby, "Prairie Fires in the North-West," *Saskatchewan History* 9 (1966): 81–91.
10 Stephen J. Pyne, *Fire in America: A Cultural History of Wildland and Rural Fire* (Princeton, NJ: Princeton University Press, 1982), 92.
11 Powell, *Report on the Lands of the Arid Region*, 195.
12 G.L. Hoxie, "How Fire Helps Forestry," *Sunset* 34 (1910): 145–51.
13 "Blames the Loggers: Secretary Ballinger Favors Compulsory Measures to Require Lumbermen to Clean Up Rubbish," *Los Angeles Times*, 25 August 1910, 12.
14 Gifford Pinchot, *Breaking New Ground* (Washington, DC: Island Press, 1998), 13.
15 John Muir, "My First Summer in the Sierra (Part IV)," *The Atlantic*, April 1911.
16 Timothy Egan, *The Big Burn* (Boston, MA: Houghton, 2009), 71.
17 "Forest Fires Make Haze, Boston Reports that Smoke from West Affects Atmosphere There," *New York Times*, 26 August 1910, 4.
18 Stephen J. Pyne, "The Source." The 2001 Distinguished Lectureship in Forest and Conservation History. Joint Conference of the American Society for Environmental History and the Forest History Society: https://foresthistory.org/wp-content/uploads/2016/12/The-Source-Stephen-Pyne-Lecture.pdf.
19 "Government Spends Millions Fighting Forest Fires," *New York Times*, 28 August 1910, SM7.
20 "Jails Give Up Prisoners to Battle with Forest Fires: Many Settlers Have Lost Everything in the Northwest. Idaho Guards May Be Set to Work – Thousands Are Fighting Fires in Coeur D'Alene Reserves – Rush Orders for Troops," *Los Angeles Times*, 14 August 1910, 7.
21 "Hundreds Missing in Forest Fires," *New York Times*, 24 August 1919, 4.
22 F.A. Silcox, "How the Fires Were Fought," *American Forestry* 16 (November 1910): 631–9.

23 "The Great Fire of 1910," *Synopsis, US Forest Service,* undated, https://www.fs.usda.gov/Internet/FSE_DOCUMENTS/stelprdb5444731.pdf.
24 "Pinchot Places Blame for Fires," *New York Times*, 27 August 1910, 3.
25 Ibid.
26 Ibid.
27 H.S. Graves, "Report of the Forester," *Annual Reports of the Department of Agriculture for the Year up until July 1913* (Washington, DC: Government Printing Office, 1913), 12.

Chapter Five

1 Rick Arthur, "Southern Alberta Wildfires of 1910," *Forestry Chronicle* 90, no. 1 (January–February 2014), 8.
2 "Forest Fires – A National Menace," *The Globe,* 8 September 1910.
3 Donald Sutherland, "Land for Settlers by Director of Colonization," *The Globe*, 10 June 1911, A7.
4 "Fires on Boundary Lines Still Raging," *Toronto Daily Star*, 9 May 1911, 3.
5 "Saving the Forests" *Toronto Daily Star*, 9 May 1911, 3.
6 Ibid.
7 R.H. Campbell, *Report of the Director of Forestry for the Year 1916* (Ottawa: J. de L. Taché, 1916), 10.
8 E.H. Finlayson, *Report of the Director of Forestry for the Year 1916* (Department of Interior, Canada), 10.
9 R.H. Campbell, *Forest Fires and Railways* (Ottawa: Government Printing Bureau, 1911).
10 Campbell, *Report of the Director of Forestry for the Year 1916*, 19.
11 Clifford Sifton, "Conservation in 1917," *The Ninth Annual Report of the Commission of Conservation* (Ottawa, 1918), 11.
12 Peter Murphy, Cordy Tymstra, and Merle Masse, "The Great Fire of 1919," *Forest History Today* (Spring–Fall 2015), 24.
13 "Lac La Biche Village in Ashes; Entire District Is Homeless; Condition of People Perilous," *Edmonton Bulletin*, 21 May 1919.
14 R.H. Campbell, "Report of the Director of Forestry," *Annual Report of the Department of the Interior for the Fiscal Year Ending March 31, 1920* (Edmonton, AB: Kings Printer 1921), 8.
15 "Will Use Airplanes to Fight Forest Fires," *The Globe*, 13 May 1921, 3.
16 H.V. Nelles, *Politics of Development: Forests, Mines, and Hydro-Electric Power in Ontario, 1849–1941*, 2nd ed. (Montreal and Kingston: McGill-Queen's University Press, 2005), 26.
17 W.R. Riddell, "Interim Report of the Commission," in *Ontario Timber Commission Interim Reports 1920* (Toronto: Clarkson W. James, 1921), https://collections.ola.org/mon/25005/15378.pdf.
18 "Minister Reduced Forest Fire Dues, Cut Them Fifty Percent," *The Globe*, 26 April 1921, 6.
19 "Forest Fires Kill Jobs," *The Globe*, 30 July 1921.
20 Edmund Zavitz, *Report of the Forestry Branch* (Toronto: Ontario Sessional Papers, 1915), 139.
21 Clifton D. Howe, "High Trees Not High Tariffs Are Key of Trade: Canada's Forests Must Be Protected," *The Globe*, 5 August 1921, 22.
22 Ibid.
23 "4,000 fires Cut Awful Swath across Dominion," *The Globe*, 6 March 1923, 6.
24 "In Former Days," *The Globe*, 6 October 1922, 4.
25 Cited in "The Great Fire of 1922," *Forestry* 4, no.1 (Spring 2013), 16.
26 "Death Toll 25 to 30; Fire Danger Now Over," *Toronto Daily Star*, 6 October 1922, 1.
27 "Supplies Are Rushed to Needy Sufferers from Forest Fires," *The Globe*, 7 October 1922, 1.
28 Peter Gillis and Thomas Roach, *Lost Initiatives: Canada's Forest Industries, Forest Policy and Conservation* (New York: Greenwood, 1986), 103–4.

Chapter Six

1 Stephen J. Pyne, *Awful Splendour: A Fire History of Canada* (Vancouver: University of British Columbia Press, 2007), 80.
2 "The Thief of The Prairie: Wolves and Coyotes the Scourge of Montana: A War of Extermination to Be Begun by the Ranchmen on the Pests of the Cattle Ranges," *New York Times*, 31 January 1887, 8.

3 William Greeley, "'Paiute Forestry' or the Fallacy of Light Burning" (1920); reprinted in *Forest History Today* 60, no. 4 (Spring 1999): 33–7.
4 Ibid.
5 A. Leopold, "'Paiute Forestry' vs. Forest Fire Prevention," *Southwestern Magazine* 2 (1920): 12–13.
6 Theodore (Ted) Binnema and Melanie Niemi, "'Let the Line Be Drawn Now': Wilderness, Conservation, and the Exclusion of Aboriginal People from Banff National Park in Canada," *Environmental History* 11, no. 4 (2006): 724–50.
7 *Annual Report of the Department of the Interior for the year 1887* (Ottawa: Maclean Roger & Company, 1888), https://babel.hathitrust.org/cgi/pt?id=nyp.33433014154433&view=1up&seq=9.
8 Binnema and Niemi, "'Let the Line Be Drawn Now,'" 736.
9 Ibid., 739.
10 Francis Parkman, *Letters of Francis Parkman*, ed. Wilbur R. Jacobs (Norman: University of Oklahoma Press, 1960).
11 Binnema and Niemi, "'Let the Line Be Drawn Now,'" 726.
12 Clyde Leavitt, "Railway Fire Protection in Canada," *The Forestry Chronicle* 4, no. 4 (1928): 10–19.
13 *Report of the Minister of Lands, Forests and Mines of the Province of Ontario* (Toronto: King's Printer, 1915).
14 A.B. Macdonald, *Department of Interior, Annual Report* (March 1911): 15–23.
15 Peter J. Murphy, Robert W. Udell, Robert E. Stevenson, and Thomas W. Peterson, *A Hard Road to Travel: Land, Forests and People in the Upper Athabasca Region* (Durham, NC: Forest History Society, 2007), 224–6.
16 Ian S.G. Puppe, "Conduits of Communion: Monstrous Affections in Algonquin Traditional Territory" (PhD dissertation, University of Western Ontario, 2015), https://ir.lib.uwo.ca/cgi/viewcontent.cgi?article=4658&context=etd.
17 R. Mawani, "Imperial Legacies (Post) Colonial Identities: Law, Space and the Making of Stanley Park, 1859–2001," *Law Text Culture* 7, no. 1 (2003), 100.
18 Jocelyn Thorpe, "To Visit and to Cut Down: Tourism, Forestry, and the Social Construction of Nature in Twentieth-Century Northeastern Ontario," *Journal of the Canadian Historical Association* 19, no. 1 (2008): 331–57.
19 Bruce W. Hodgins and Jamie Benidickson, "Resource Management Conflict in the Temagami Forest, 1898 to 1914," *Historical Papers, London* 13, no. 1 (1978), 158.
20 Patricia McCormack, "The Political Economy of Bison Management in Wood Buffalo National Park," *Arctic* 45, no. 4 (December 1992): 367–80.
21 John Macoun of the Geological Survey of Canada (1882), quoted in F.R. Holt, *Out of the Flames: Fires and Fire Fighting on the Canadian Prairies* (Calgary, AB: Fifth House, 1998), 21–2.
22 R.H. Campbell, Department of Interior directive, *Annual Report of the Department of Indian Affairs* (Ottawa. Printer to the Queen's Most Excellent Majesty, 1895).
23 John Sandlos, "Not Wanted in the Boundary: The Expulsion of the Keeseekowenin Ojibway Band from Riding Mountain National Park," *Canadian Historical Review* 89, no. 2 (June 2008): 189–221.
24 Ibid., 10.

Chapter Seven

1 Brant Irving. "Forest Lookouts Defy Lightning," *Scientific American* 138, no. 6 (1928): 489–92. http://www.jstor.org/stable/24964982.
2 Arthur Lower, *My First Seventy-five Years* (Toronto: MacMillan, 1967), 64.
3 Ibid.
4 Rob Corcoran, "The 100th Anniversary of Bush Flying in Canada," *Temagami Times*, Summer 2019, 1,3.
5 "Government of Ontario, Curtiss HS2L H-boat," Canadian Bush Plane Heritage Centre, https://www.bushplane.com/opas-aircraft/histories-opas-hboat/.
6 "Forest Fire Damage Greatest in History," *Los Angeles Times*, 10 October 1919.
7 Robert W. Cermak, "Pioneering Aerial Forest Fire Control: The Army Air Patrol in California, 1919–1921," *California History* 70, no. 3 (Fall 1991): 290–305.

8 "Air Patrols Prevent State Forest Fires: Seven More Are Ordered for California National Reserves," *Los Angeles Times*, 22 August 1919, 6.
9 In April 1915, Captain J.M. Furnival was the first person on the ground to hear the voice of a pilot. Major Prince sent word: "If you can hear me now, it will be the first time speech has ever been communicated to an airplane in flight."
10 William B. Crist and Bruce Mcleod, "Birds of War," *MacLean's Magazine* (15 December 1943): 20–38.
11 Diane M. Smith, "Sustainability and Wildland Fire, the Origins of Forest Service Wildland Fire Research," US Forest Service, United States Department of Agriculture (2017), 34, https://www.fs.usda.gov/treesearch/pubs/54271.
12 Richard Elsom, "T.V. Pearson and the Parachute Scheme," *The Text Message*, Blog of the Textual Records Division of the National Archives, 29 August 2017, https://text-message.blogs.archives.gov/2017/08/29/t-v-pearson-and-the-parachute-scheme.
13 Richard Elsom, "T.V. Pearson and The Parachute Scheme," *Smokejumper Magazine* (16 January 2018), 9.

Chapter Eight

1 James J. McEntee, *Now They Are Men*, ed. Sherman F. Mittell (Washington, DC: National Home Library Foundation, 1940), 61.
2 United States Congress, *Department of Labor-Federal Security Agency Appropriations Bill for 1943, Hearings of the Committee on Appropriations House of Representatives* (Washington, DC: United States Government Printing Office, 1942), 957.
3 "60 Forests Protected by the CCC" (Washington, 1938), 1– 3; C.R.M., No. 788, CCC in Emergencies. 4.
4 "Forest Fire Fatal to 14 Is Dying Out: 4-Day Battle against Blaze in Wyoming Mountains Is Close to Victory. Human Blame Is Denied. Reservation Chief Attributes Casualties to Shifting Wind of 'Gale-Like' Force. Many Need Only First Aid. Ground Too Hot to Cross. Dead Ranger Once Lived Here," *New York Times*, 24 August 1937.
5 "2,600,000 Had Jobs in CCC in Six Years: Fechner's Posthumous Report Shows 1,855,000,000 Trees Planted, Vast Land Work Benefit to Nation Cited. Late Director Stressed Saving of Man Power and Defense Gain by Youth Training Use of Manpower. Stressed Conservation Work Widespread," *New York Times*, 15 January 1940.
6 63. United States Congress, Senate, Appropriations Committee, *Labor-Federal Security Appropriation Bill for 1943, Hearings before the States Senate, Seventy-Seventh Congress ending June 30, 1943*.
7 Ibid.
8 *U.S. Government Campaign for Wartime Forest Fire Prevention for 1943*, prepared for U.S. Department of Agriculture Forest Service in co-operation with the Bureau of Campaigns. Office of War Information, 1943.
9 Ibid., 965.
10 "Operation Firefly & the 555th," https://www.fs.usda.gov/science-technology/fire/people-working-fire/smokejumpers/smokejumper-base-contact-information/missoula-smokejumpers/history/operation.
11 United States Office of War Information, *Victory, the Official Weekly Bulletin of the Office of War Information* (3 May 1943): 266.

Chapter Nine

1 Desmond Morton, *A Military History of Canada: From Champlain to Kosovo* (Toronto: McClelland and Stewart, 1985), 175.
2 "Safe and Sound Forestry," *The Globe*, 21 August 1933, 4.
3 Thomas R. Roach and Richard Judd, "A Man for All Seasons: Frank John Dixie Barnjum, Conservationist, Pulpwood Embargoist and Speculator!," *Acadiensis* 20, no. 2 (1991): 129–44.
4 F.J.D. Barnjum, "Forest Fire Statistics," *The Globe*, 4 July 1932.
5 L. Anders Sandberg and Peter Clancy, "Forestry in a Staples Economy: The Checkered Career of Otto Schierbeck, Chief Forester, Nova Scotia, Canada, 1926–1933," *Environmental History* 2, no. 1 (1997): 74–95.

6 Ibid.
7 Barnjum, "Forest Fire Statistics," 2.
8 E.H. Finlayson to W.W. Cory, 30 January 1923. RG 39, Records of the Dominion Forest Service, vol. 44 file 12–13, National Archives of Canada.
9 Ibid, 144.
10 *Saturday Night*, 24 November 1923.
11 F.C. Mears, "Aggressive Policy Needed for Forests: Pulpwood Commission Urges Active Measures of Conservation in Canada. Embargo Not Favoured," *The Globe*, 6 September 1924, 3.
12 Frank Barnjum, "Plans for Forest Reforms," *The Globe,* 23 January 1933, 4.
13 "A Crusader Called Away," *The Globe,* 21 February 1933, 4.
14 Special Cable to the *New York Times*. "Body of Barnjum to Be Brought Here: Canadian Forest Conservationist, Who Died in Paris, Long Prominent in Paper Trade," *New York Times*, 22 February 1933.
15 "Millionaire Lumberman Dies," *Wall Street Journal*, 21 February 1933, 8.
16 "Destructive Forest Fires," *The Globe*, 4 June 1935, 4.
17 *Proceedings of the Fifth Pacific Science Congress: Victoria, B.C., June 1–4, Vancouver, B.C., June 5–14, Post-congress Tour, June 5–19, Canada, 1933*. National Research Council of Canada, 19 October 2021.
18 "Making Whoopee up North," *The Globe*, 17 April 1934, 6.
19 Trudy Copithorne, "The Fire of 1936," *Chaps and Chinooks: A History West of Calgary, Vol. 1*, ed. Evelyn Buckley (Calgary, AB: Foothills Historical Society, 1976).
20 Copithorne, "The Fire of 1936."
21 "Forest Fire Pall Fools Birds and Tots," *The Globe*, 1 August 1936, 6.
22 John R. Taylor versus The London Assurance Corporation, Supreme Court of Canada, 1935, https://scc-csc.lexum.com/scc-csc/scc-csc/en/3539/1/document.do.
23 C.W. Irwin, "Astounding Toll of Forest Fires," *Globe and Mail*, 9 June 1937, 6.
24 Irwin, "Astounding Toll of Forest Fires," 1.
25 "Mad Convict Blamed for Forest Fires," *The Globe*, 17 July 1936, 9.
26 Holroyd, "Boundary Creek Fire," *Partial Yearbook for 1935 (Aug 1 to Oct 1)*, https://parkwardenalumni.com/wp/wp-content/uploads/2020/06/Boundary-Creek-Fire-Holroyd-diary-1935.pdf.
27 John Parminter, "The Merville Fire of 1922," *British Columbia Forest History Newsletter* (Winter 2018): 1– 6.
28 John Parminter, "Darkness at Noon – The Bloedel Fire of 1938," *Research Branch B.C. Forest Service* (October 1994), https://www.for.gov.bc.ca/hfd/pubs/docs/scv/scv871.pdf.
29 From Our Correspondent, "Forest Fires Less Threatening," *Times*, 26 July 1938, 12. *The Times Digital Archive*, link-gale-com.proxy.queensu.ca/apps/doc/CS202978042/TTDA?u=queensulaw&sid=bookmark-TTDA&xid=b6f13724.
30 From Our Correspondent, "Forest Fires Less Threatening," 12.
31 "C.C.C. Lauded by Forester," *Los Angeles Times*, 10 February 1938, 3.
32 Quoted in Parminter, "Darkness at Noon," 38.
33 "Again the Forest Fire Horror," *Globe and Mail*, 12 October 1938, 6.
34 "Premier Sails for Bermuda," *Globe and Mail*, 12 October 1938.
35 "Says Ontario Shuns Experts in Forestry," *Globe and Mail*, 2 November 1938, 4.
36 Ibid.
37 John C.W. Irwin, "The Canadian Forest and Its Future," *The Annals of the American Academy of Political and Social Science* 253 (September 1947): 52– 8.
38 Wellington Jeffers, "Extravagant and Wasteful Depletion of Canada's Forest Resources Will Mean Many Lost Jobs Ten Years from Now," *Globe and Mail*, 9 December 1938, 22.

Chapter Ten

1 "The Not So Mysterious Disappearance of E.H. Finlayson," *Ottawa Citizen*, 28 February 2001, 1.
2 I.C.M. Place, "Forestry Canada Celebrates 75 Years of Research," *The Forestry Chronicle* 69, no. 3 (June 1993), 273.

3 Gabrielle Blais, "Recovering a Lost Heritage: The Case of the Canadian Forestry Service Records," *Archivaria* 25 (Winter 1987): 84.
4 John Bacher, *Two Billion Trees and Counting, The Legacy of Edmund Zavitz* (Toronto: Dundurn, 2011), 187–8.
5 Peter Gillis and Thomas R. Roach, *Lost Initiatives: Canada's Forest Industries, Forest Policy and Forest Conservation*, Contributions in Economics and Economic History, no. 69 (New York: Greenwood, 1986), 225.
6 Neil McKenty, *Mitch Hepburn* (Toronto: McClelland and Stewart, 1967).
7 "Canada Completes Study of Forest Fire Dangers," *New York Times*, 28 July 1935, 27.
8 Stephen J. Pyne, *Awful Splendour: A Fire History of Canada* (Vancouver: University of British Columbia Press, 2007), 266.
9 N. Winston, "Annual Cost of Canadian Corps Plan Set at $54,000,000 by Experts," *Globe and Mail*, 27 January 1939, 15.

Chapter Eleven

1 John C.W. Irwin, "Ontario Sort of Lookout Towers," *Globe and Mail*, 2 April 1945, 6.
2 John C.W. Irwin, "New Deal for Forests Would Save Wealth in Ontario," *Globe and Mail*, 6 April 1945, 6.
3 Gordon Hak, *Capital and Labour in the British Columbia Forest Industry, 1934–74* (Vancouver: University of British Columbia Press, 2006), 50.
4 R. Peter Gillis and Thomas R. Roach, *Lost Initiatives: Canada's Forest Industries, Forest Policy and Forest Conservation* (New York: Greenwood, 1986), 66.
5 "Forests League Calls for Control by Commission," *Globe and Mail*, 16 October 1945, 4.
6 Cited in Douglas White Ambridge, "Fears and Fancies of the Forests," (24 March 1955), in *The Empire Club of Canada Addresses* (Toronto), 233–42, https://speeches.empireclub.org/details.asp?ID=60911.
7 Ibid.

Chapter Twelve

1 "Women in Towers Better than Men," *Globe and Mail*, 18 August 1937, 11.
2 Ibid.
3 "Taking Her Two Tots in Canoe, Ranger's Wife Quells Bush Fire," *Globe and Mail*, 5 October 1950, 17.
4 "Forest Fires Laid to Women," *New York Times*, 13 September 1936, E7.
5 "Enroll Indians as Firefighters," *The Globe*, 24 June 1911, 26.
6 James G. Lewis and Robert Hendricks, "A Brief History of African Americans and Forests," Forest Service, https://www.fs.fed.us/people/aasg/PDFs/African_Americans_and_forests_March21%202006.pdf.
7 Tanya Lee Stone, *Courage Has No Color: The True Story of the Triple Nickles, America's First Black Paratroopers* (New York: Penguin Random House, 2013).
8 "Giving Voice to Cultural Safety of Indigenous Wildland Firefighters in Canada," Turtle Island Consulting Services Inc., North Saanich BC, 20 October 2021.
9 Emily Stifler Wolfe, "Women in Fire, Fuel, Oxygen and Heat," *Mountain Outlaw Magazine* (April 2018), https://www.iawfonline.org/article/women-in-fire-fuel-oxygen-and-heat/.
10 Ibid.

Chapter Thirteen

1 "Plan Conscripting Men to Combat Forest Fires," *Globe and Mail*, 24 June 1947, 6.
2 Joyce Butler, *Wildfire Loose: The Week Maine Burned* (Kennebunkport, ME: Down East Books, 1997), 122.
3 "Join to Fight Forest Fires," *New York Times*, 28 April 1937, 11.
4 Northeastern Forest Fire Protection Commission website, https://www.nffpc.org/en/information/about.
5 "Forest 'Fire' Loses to 7-State Defense: An Interstate Cooperative Movement to Fight Forest Fires," *New York Times*, 6 October 1951, 23.

6 Jack Piner, "Goal Is to Develop Virus to Kill off Forest Pests," *Globe and Mail*, 27 September 1950, 17.
7 Stephen J. Pyne, *Awful Splendour: A Fire History of Canada* (Vancouver: University of British Columbia Press, 2007), 338.

Chapter Fourteen

1 Helen Sawyer Hogg, "Blue Sun," *Journal of the Royal Astronomical Society of Canada* 44, (December 1950): 241.
2 "Huge Area Dimmed by Smog from Canada Forest Fires," *Hartford Courant*, 25 September 1950, 1.
3 N.P. Carlson, "One Sunday Afternoon in 1950, the Sun Went Out," *Post-Journal*, Jamestown, NY, 3 January 1987.
4 "Forest Fires Cast a Pall on Northeast, Canadian Drift 600 Miles Long Darkens Wide Areas and Arouses Atom Fears," *New York Times*, 25 September 1950.
5 Ibid.
6 Ibid.
7 Carlson, "One Sunday Afternoon in 1950, the Sun Went Out."
8 Cordy Tymstra, *The Chinchaga Firestorm* (Edmonton: University of Alberta Press, 2015), 44–6.
9 "Blue Tint to the Sun, Haze at 30,000 Feet," *The Times*, 27 September 1950, 1.
10 "Toronto Sky Darkened by Smoke in Alberta, *Globe and Mail*, 25 September 1950, 1.
11 Ibid.
12 "Devil Has Moved in with Us, Firefighting Farmers Moan," *Globe and Mail*, 28 September 1950, 17.

Chapter Fifteen

1 R.P. Turco, O.B. Boon, T.P Ackerman, et al., "Nuclear Winter: Global Consequences of Multiple Nuclear Explosions," *Science* 222, no. 4630 (December 1983): 1283–92.
2 Paul Crutzen and John Birks, "Twilight at Noon: The Atmosphere after a Nuclear War," *Ambio* 11, no. 2–3 (1982): 114–25.
3 Paul R. Ehrlich, John Harte, Mark A. Harwell, et al., "Long-term Biological Consequences of Nuclear War," *Science* 222, no. 4630 (December 1983): 1293–300.
4 "Controlled Forest Fire in Ontario to Test 'Nuclear Winter' Theory," *Toronto Star*, 15 July 1985, A2.
5 Harrison E. Salisbury, *The Great Black Dragon Fire: A Chinese Inferno* (Boston: Little, Brown, 1989), 11.

Chapter Sixteen

1 H.H. Chapman, "Forest Fires and Forestry in the Southern States," *American Forests* 18, 510–17.
2 Hal K. Rothman, *Blazing Heritage: A History of Wildland Fire in the National Parks* (New York: Oxford University Press, 2007), 60.
3 For insights into both essays, see Andrew Larson, "Introduction to the Article by Elers Koch, 'The Passing of the Lolo Trail,'" *Fire Ecology* 12, no. 1 (2016): 7–12.
4 Report T-169, *Fire History and Fire Records for Everglades National Park, 1948–49*, Everglades National Park, South Florida Research Center, 47.
5 Ibid., 48.
6 Rothman, *Blazing Heritage*, 85.
7 Timothy Egan, "Nature and Goal of Guarding Land Served as Fuel in Yellowstone's Fires," *New York Times*, 22 September 1988, 1.
8 Henry Adams, Raymond Burns Russel, B.P. Chuck Powell, and Carmela Schmidt, "Letters to the *Times*: Yellowstone Park Fires," *Los Angeles Times*, 18 September 1988, 5.
9 Philip Shabecoff, "Park and Forest Service Chiefs Assailed on Fire Policy," *New York Times*, 10 September 1988, 6.
10 Ibid.
11 Ibid.
12 Jerry T. Williams, "Managing Risk in Wilderness Fire Management," in *Proceedings: Symposium on Fire in Wilderness and Park Management*, ed. James K. Brown, Robert W. Mutch, Charles W. Spoon, and Ronald H. Wakimoto (Ogden, UT: US Forest Service, Intermountain Research Station, 1995), 22.

13 Bob Specter and Tamara Jones, "Yellowstone: Natural Lab or Business?," *Los Angeles Times*, 16 September 1988, B1.
14 M. Rupert Cutler, "Fire in Yellowstone, Hot Air in D.C.," *New York Times*, 11 September 1988, E31.
15 Steven Hostetler et al., *Greater Yellowstone Climate Change Assessment, Past, Present, and Future Climate Change in Greater Yellowstone Watersheds* (Boseman: Montana State University, 2021).

Chapter Seventeen

1 Scott E. Nielsen, Marc R.L. Cattet, John Boulanger, et al., "Environmental, Biological and Anthropogenic Effects on Grizzly Bear Body Size: Temporal and Spatial Considerations," BMC *Ecology* 13, no. 31 (2013), 13–31.
2 US Fish and Wildlife Service, Mountain-Prairie Region, "Wildfires and Grizzly Bears" (June 2003), https://www.fws.gov/mountain-prairie/species/mammals/grizzly/wildfire&bears.pdf.

Chapter Eighteen

1 Geoffrey York, "Smoke from 224 Forest Fires Threatens Manitoba Airports," *Globe and Mail*, 25 July 1989, 1.
2 Kevin Hirsch, "A Chronological Overview of the 1989 Fire Season in Manitoba," *The Forestry Chronicle* 67, no. 4 (1991), 358–65.
3 Ibid.
4 Don MacIvor, forest meteorologist with the Federal Atmospheric Environment Service, quoted in "How to Set a Forest Fire," *Globe and Mail*, 31 July 1989.
5 Hirsch, "A Chronological Overview of the 1989 Fire Season in Manitoba," 360.
6 Diane M. Smith, "Yellowstone National Park and the Summer of Fire," *Rural Connections* 7, no. 1: 31–4, https://www.fs.usda.gov/treesearch/pubs/43888.
7 Brian J. Stocks, "Federal Forest Fire Research in Canada: An Impressive Past, a Troubled Present, and an Uncertain Future," presentation to Wildland Fire Canada Conference (Banff, Alberta, 2012).
8 "Scientists Worried by Sharp Rise in Damage Caused by Forest Fires," *Globe and Mail*, 28 July 1989, A11.
9 B. Harrington, "Climate Changes, A Review of Causes," *Canadian Journal of Forestry Research* 17, no. 11 (1987): 1313–39.
10 C.E. Van Wagner, "The Historical Pattern of Area Burned in Canada," *Forest Chronicle* 64 (June 1988): 182, 185.
11 M.D. Flannigan and C.E. Van Wagner, "Climate Change and Wildfire in Canada," *Canadian Journal of Forest Research* 21 (1991): 66–72.
12 Michael Weber and Brian J. Stocks, "Forest Fires and Sustainability in the Boreal Forests of Canada," *Ambio* 27, no. 7 (1998): 545–50.

Chapter Nineteen

1 Graeme Worboys, "A Brief Report on the 2003 Australian Alps Bushfires," *Mountain Research and Development* 23, no. 3 (2003) 294–5.
2 Maria Carolina Varela, "The Deep Roots of the 2003 Forest Fires in Portugal," *International Forest Fire News* 34 (January–June 2006): 2–22.
3 S.J. Colombo, M.L. Cherry, C. Graham, et al., "The Impacts of Climate Change on Ontario's Forests," *Forest Research Information Paper* no. 143, Sault Ste Marie, Ontario: Ontario Forest Research Institute, 1998, https://fgca.net/wp-content/uploads/OFRI-impact-of-climate-change-on-ontarios-forests-1.pdf.
4 Elisabeth Bumiller, "Bush Flies Over Fire Site to Promote 'Healthy Forest,'" *New York Times*, 12 August 2003, A10.
5 US Forest Service, Department of Agriculture, *The 2003 San Diego County Fire Siege Fire Safety Review* (2004).
6 Based on various data supplied by the National Interagency Fire Center.
7 Government of Alberta, "Final Documentation Report, Chisholm Fire," LWF- 0630 (2001).
8 Efrain Hernandez Jr, "Fire Unit Helps Gang Members Restart Lives: Forest Service: Fighting Wilderness Blazes Keeps Young Adults off the Streets but the Work Is Only Seasonal," *Los Angeles Times*, 27 November 1994, 8.

9 Gil Reavill, "Meltdown in Los Alamos," *Maxim* (October 2000), 1.
10 Thomas Zimmerman, Laurie Kurth, and Mitchell Burgard, "The Howling Prescribed Natural Fire – Long-term Effects on the Modernization of Planning and Implementation of Wildland Fire Management," *Proceedings of third Fire Behavior and Fuels Conference 25–29 October 2010, Spokane, WA.* (Birmingham, AL: International Association of Wildland Fire, 2011).
11 Mark Heathcott, "Notes on 2003 Fire Season," *Parks Canada,* Calgary, Spring 2003. Courtesy of Mark Heathcott
12 Ed Struzik, "Underground Intelligence, the Need to Map, Monitor, and Manage Canada's Groundwater Resources in an Era of Drought and Climate Change," presentation at the Program on Water Issues, Munk School of Global Affairs, University of Toronto, 11 June 2013.
13 "Top Ten Weather Stories of 2002," Environment Canada, 8 August 2017, http://www.ec.gc.ca/meteo-weather/default.asp?lang=En&n=7E0A416A-1#t1. Environment Canada.
14 Author's notes. Interview with Langdon, 1 September 2003.
15 These comments were all recorded by the author when he was in the field with firefighters in 2003.
16 Volker C. Radeloff, David P. Helmers, H. Anu Kramer, et al., "Rapid Growth of the US Wildland-urban Interface Raises Wildfire Risk," *Proceedings of the National Academy of Sciences* 115, no. 13 (March 2018): 3314–19.
17 City of Kelowna, British Columbia, "Compact, Complete Communities," *Facts in Focus* (newsletter, 2018).
18 John C.W. Irwin, "The Canadian Forest and Its Future," *Annals of the American Academy of Political and Social Science* 253, Features of Present-Day Canada (September 1947): 52–8.
19 Gary Filmon, "Firestorm 2003: Provincial Review," Government of British Columbia, 15 February 2004.
20 M.D. Flannigan, K.A. Logan, B.D. Amiro, et al., "Future Area Burned in Canada," *Climatic Change* 72 (January 2005): 1–16.
21 Christopher Adams, AScT, Oil and Gas Specialist, "Summary of Shale Gas Activity in Northeast British Columbia 2011," Government of British Columbia, Ministry of Energy and Mines Geoscience and Strategic Initiatives Branch.
22 The Canada West Foundation, "State of the West Report" (2008).

Chapter Twenty

1 *May 2016 Wood Buffalo Wildfire, Post-Incident Assessment Report,* KPMG, prepared for Alberta Emergency Management Agency, Final Report, May 2017.

Chapter Twenty-One

1 Cordy Tymstra, *The Chinchaga Firestorm* (Edmonton: University of Alberta Press, 2015).
2 John Thistleston, "Researchers Confirm First Fire Tornado during 2003 Bushfires," *Sydney Morning Herald*, 19 November 2012, 1.
3 David A. Peterson, Edward J. Hyer, James R. Campbell, et al., "The 2013 Rim Fire: Implications for Predicting Extreme Fire Spread, Pyroconvection, and Smoke Emissions," *Bulletin of the American Meteorological Society* 96, 2 (2015): 229–47.
4 Andrew J. Dowdy, Michael D. Fromm, and Nicholas McCarthy, "Pyrocumulonimbus Lightning and Fire Ignition on Black Saturday in Southeast Australia," *Journal of Geophysical Research: Atmospheres* 122 (2017): 7342– 54.
5 N.P. Lareau, N.J. Nauslar, and J.T. Abatzoglou, "The Carr Fire Vortex: A Case of Pyrotornadogenesis?" *Geophysical Research Letters* 45, no. 13 (2018): 107–13.
6 David A. Peterson et al., "Wildfire-driven Thunderstorms Cause a Volcano-like Stratospheric Injection of Smoke," *Nature, npj Climate and Atmospheric Science* 1, no. 30 (2018), https://doi.org/10.1038/s41612-018-0039-3.
7 J.T. Fasullo, "Coupled Responses to Recent Australian Wildfire and COVID-19 Emissions Anomalies Estimated in CESM2," *Geophysical*

Research Letters 48, no. 15 (27 July 2021), https://doi.org/10.1029/2021GL093841.

Chapter Twenty-Two

1 "Top 10 Warmest Years on Record," *Climate Central* (15 January 2020).
2 Bill Gabbert, "Rim Fire Burn Area – 'Nuked' or Not?" *Wildfire Today* (19 September 2003): 3.
3 Joseph W. Wagenbrenner, Lee H. MacDonald, Robert N. Coats, et al., "Effects of Post-fire Salvage Logging and a Skid Trail Treatment on Ground Cover, Soils, and Sediment Production in the Interior Western United States," *Forest Ecology and Management* 335 (2015): 176–93.
4 Kimberley T. Davis, Solomon Z. Dobrowski, Philip E. Higuera, et al., "Wildfires and Climate Change Push Low-elevation Forests across a Critical Climate Threshold for Tree Regeneration," *Proceedings of the National Academy of Sciences* 116, no.13 (March 2019): 6193–8.

Chapter Twenty-Three

1 S.H. Cannon, P. Powers, and W. Savage, "Fire-related Hyper Concentrated and Debris Flows on Storm King Mountain, Glenwood Springs, Colorado, USA," *Environmental Geology* 35, no. 2–3 (1998): 210–18.
2 Ricardo Baca, "Platte River Bounces Back for Fly Fishing after Hayman Fire," *The Denver Post*, 14 May 2012, 1.
3 Charles C. Rhoades, Alex T. Chow, Timothy P. Covino, et al., "The Legacy of a Severe Wildfire on Stream Nitrogen and Carbon in Headwater Catchments," *Ecosystems* 22 (2019): 643–57.
4 T. Gartner, J. Mulligan, R. Schmidt, and J. Gunn, *Natural Infrastructure: Investing in Forested Landscapes for Source Water Protection in the United States, Report of the World Resources Institute* (Washington, DC: World Resources Institute, 15 October 2013).
5 H.G. Smith, G.J Sheridan, P.N.J. Lane, P. Nyman, and S. Shane Haydon, "Wildfire Effects on Water Quality in Forest Catchments: A Review with Implications for Water Supply," *Journal of Hydrology* 396, nos.1–2 (2011): 170–92.
6 Personal interview. More detail can be found in D.A. Martin, "At the Nexus of Fire, Water and Society," *Philosophical Transactions of the Royal Society B: Biological Sciences*, 2016, https://doi.org/10.1098/rstb.2015.0172.
7 "Canada's Forest-related Contributions to the Sustainable Development Goals under Review in 2018," United Nations Forum on Forests as a Contribution towards the 2018 Review of the SDGs, https://www.un.org/esa/forests/wp-content/uploads/2018/03/Input_SDGs2018_Canada.pdf.
8 John A, Moody and Deborah A. Martin, "Initial Hydrologic and Geomorphic Response Following a Wildfire in the Colorado Front Range," *Earth Surface Processes and Landforms* 26, no. 10 (September 2001) 1049–70.
9 Erin N. Kelly, David W. Schindler, Vincent L. St Louis, et al., "Forest Fire Increases Mercury Accumulation by Fishes via Food Web Restructuring and Increased Mercury Inputs," *Proceedings of the National Academy of Sciences* 103, no. 51 (December 2006): 19380–5.
10 G. Wiener, B.C. Knights, M.B. Sandheinrich, et al., "Mercury in Soils, Lakes and Fish in Voyageurs National Park," *Environmental Science & Technology* 40, no. 20 (2006): 6261–8.
11 "Mercury in the Arctic," Arctic Council on-line publication, Dec 2015, https://arctic-council.org/en/news/mercury-in-the-arctic/.
12 CalFire, Government of California website, https://www.fire.ca.gov/stats-events/.
13 "WWF-Canada's 2020 Watershed Report, A National Assessment of Canada's Freshwater," (Toronto: World Wildlife Fund Canada, 2020), https://wwf.ca/wp-content/uploads/2020/10/WWF-Watershed-Report-2020-FINAL-WEB.pdf.

Chapter Twenty-Four

1 Copernicus Atmosphere Monitoring Service (CAMS), "CAMS Monitors Unprecedented Wildfires in the Arctic," blog post, 11 July 2019,

https://atmosphere.copernicus.eu/cams-monitors-unprecedented-wildfires-arctic.

2 Brigit Braune et al., "Mercury in the Marine Environment of the Canadian Arctic: Review of Recent Findings," *Science of the Total Environment* 509–10, 15 March 2015, 67–90.

3 Ibid.

4 Erik Lutsch, Kimberly Strong, Dylan B.A. Jones, et al., "Unprecedented NH_3 Emissions Detected in the High-Arctic from the 2017 Canadian Wildfires," Presentation at the Advancing Earth and Space Science (AGU) Fall Meeting, December 2018, Washington, DC.

5 W.D. Hansen and M.G. Turner, "Origins of Abrupt Change: Postfire Subalpine Conifer Regeneration Declines Nonlinearly with Warming and Drying," *Ecological Monographs* 89, no 1 (2019): 8178–202.

6 Anastasia Sniderhan and Jennifer L. Baltzer, "Growth Dynamics of Black Spruce (*Picea mariana*) in a Rapidly Thawing Discontinuous Permafrost Peatland: Growth Dynamics Boreal Peatlands," *Journal of Geophysical Research: Biogeosciences* (2016): 121.

7 J.E.D. Miller, H.T. Root, and H.D. Safford, "Altered Fire Regimes Cause Long-term Lichen Diversity Losses," *Global Change Biology* 24 (2018): 4909–18.

8 Randi Jandt et al., "Slow Recovery of Lichen on Burned Caribou Winter Range in Alaska Tundra: Potential Influences of Climate Warming and Other Disturbance Factors," *Arctic, Antarctic, and Alpine Research* 40, no. 1 (2008): 89–95.

9 Jason Adams, "Yukon Fire History Metadata," Yukon Fire History GIS Dataset 1946–2004, https://gsg.uottawa.ca/data/open/svgmetadata/yukonfirehistory2004metadata.pdf.

10 Torbjörn Johnsen, "Forest Fires in Sweden – Huge Areas Burned in 2018," editorial blog post, *Forestry*, https://www.forestry.com/editorial/forest-fires-sweden/.

11 "Lightning Triggers Fires across Alaska," Earth Observatory, EOS Project Science Office, http://earthobservatory.nasa.gov/IOTD/view.php?id=4591http://earthobservatory.nasa.gov/IOTD/view.php?id=4591.

Chapter Twenty-Five

1 "Smoke from Siberian Wildfires," National Aeronautics and Space Administration, MODIS, 6 August 2021, https://modis.gsfc.nasa.gov/gallery/individual.php?db_date=2021-08-07.

2 G.P. Kablick, D.R. Allen, M.D. Fromm, and G.E. Nedoluha, "Australian PyroCb Smoke Generates Synoptic-scale Stratospheric Anticyclones," *Geophysical Research Letters* 47 (2020), 1–9.

3 A.G. Rappold, W.E. Cascio, V.J. Kilaru, et al., "Cardio-respiratory Outcomes Associated with Exposure to Wildfire Smoke Are Modified by Measures of Community Health," *Environmental Health* 11, no. 71 (2012): 11–71.

4 Paul Rogers, "Smoke from California Fires May Have Killed More than 1,000 People. Stanford Study Highlights the Danger of Breathing High Levels of Soot," *Mercury News*, 23 September 2020.

5 Xiaodan Zhou, Kevin Josey, Leila Kamareddine, et al., "Excess of COVID-19 Cases and Deaths Due to Fine Particulate Matter Exposure during the 2020 Wildfires in the United States," *Science Advances* 7, no. 33 (13 August 2021), https://www.ncbi.nlm.nih.gov/pmc/articles/PMC8363139/.

6 S.M. Holm, M.D. Mille, and J.R. Balmes, "Health Effects of Wildfire Smoke in Children and Public Health Tools: A Narrative Review," *Journal of Exposure Science & Environmental Epidemiology* 31, no. 1 (2021): 1–20.

7 C.L. Reddington, Luke Coniber, Suzanne Robinson, "Air Pollution from Forest and Vegetation Fires in Southeast Asia Disproportionately Impacts the Poor," *GeoHealth* 5 (2021): 1–20.

8 Gary A. Morris, Scott Hersey, Anne M. Thompson, et al., "Alaskan and Canadian Forest Fires Exacerbate Ozone Pollution over Houston, Texas, on 19 and 20 July 2004," *Journal of Geophysical Research: Atmospheres* 111, no. 24 (2006), 1–10.

9 Russell W. Long, Andrew Whitehill, Andrew Habel, et al., "Comparison of Ozone Measurement Methods in Biomass Burning Smoke: An Evaluation under Field and Laboratory

Conditions," *Atmospheric Measurement Techniques* 14, no. 3 (2021): 1783–800.

10 Cheryl Dykstra et al., "Trends and Patterns of PCB, DDE and Mercury in Bald Eagle Nestlings in the Upper Midwest," *Journal of Great Lakes Research* 42, no. 2 (April 2019): 252–62.

11 Action Memorandum Amendment, United States Environmental Protection Agency Region 8. Denver, 28 August 2012, 5, https://semspub.epa.gov/work/08/1242675.pdf.

12 Nikia Hernandez, email to Christina Progress, 7 July 2016, https://semspub.epa.gov/work/08/1772063.pdf.

13 Jane Braxton Little, "Forest Fires Are Setting Chernobyl's Radiation Free," *The Atlantic* (10 August 2020), https://www.theatlantic.com/science/archive/2020/08/chernobyl-fires/615067/.

14 Alan A. Ager, Richard L. Lasko, Viktor Myroniuk, et al., "The Wildfire Problem in Areas Contaminated by the Chernobyl Disaster," *Science of the Total Environment* (2019), 696.

15 Rosana Aguilera, Thomas Corringham, Alexander Gershunov, et al., "Fine Particles in Wildfire Smoke and Pediatric Respiratory Health," *Califormia Pediatrics* (April 2021), https://pubmed.ncbi.nlm.nih.gov/33757996/.

Chapter Twenty-Six

1 "A Forest of Fire," *New York Times*, 11 September 1870, 4.

2 "Forest Fires," *New York Times*, 12 November 1874, 3.

3 "Wisconsin Forest Fires: Homeless Families Number 700 – Heavy Damage to Timber," *New York Times*, 16 August 1870, 10.

4 "The Forestry Commission," *New York Times*, 19 February 1886, 6.

5 "Pennsylvania Forest Fires," *New York Times*, 29 April 1871, 3.

6 "Endangered by Forest Fires," *New York Times*, 22 August 1883, 5.

7 "Forest Fire Raging," *New York Times*, 26 September 1886, 2.

8 "Forest Fire Unchecked," *New York Times*, 20 August 1895, 11.

9 Gifford Pinchot, *The Fight for Conservation* (Project Gutenberg Ebook, 23 February 2004), https://www.gutenberg.org/ebooks/11238.

10 Stephen Ponder, "Gifford Pinchot, Press Agent for Forestry," *Forest & Conservation History* 31, no. 1 (January 1987): 26–35.

11 Pinchot to R.C Melward, 20 May 1903, Office of Reserves Correspondent, Record Group 95, National Archives.

12 Ponder, "Gifford Pinchot, Press Agent for Forestry," 28.

13 "Gifford Pinchot's Life Has Been All Battles: Never a Machine Man, Pennsylvanian Has Won Out in Politics for 20 Years by His Own Shrewdness," *Boston Globe*, (25 May 1930), 6.

14 "Pinchot for President," *West Point Times Herald*, cited in the *Washington Post*, 11 May 1906, 6.

15 "Severe Criticism, A Faddist, A Statesman and an Egoist," *L.A. Times*, 3 August 1909, I17.

16 "Remarks by Staff, the Fizzles Fizzled," *L.A Times*, 16 May 1910, P16.

17 George Kibbe Turner, "The New Art of Making Presidents," *McClure's Magazine* (July 1912): 317–27.

18 "Pinchot vs Carter," *New York Times*, 2 September 1910, 8.

19 "Pinchot Places Blame for Fires," *New York Times*, 27 August 1910, 3.

20 "Saving the Forest," *Toronto Daily Star*, 14 July 1911, 8

21 Stephen J. Pyne, *Fire in America: A Cultural History of Wildland and Rural Fire* (Princeton, NJ: Princeton University Press, 1982), 272.

22 "The Fire-Swept North," *The Globe*, 13 July 1911, 6.

23 "Launch Towed Scow with the Bodies of Dead to Lonely Cemetery," *Toronto Daily Star*, 17 July 1911, 2.

24 James G. Lewis, "Time, Fire, and Taxes: Frederick E. Weyerhaeuser at the American Forest Congress," *Forest History Today* (Spring–Fall 2005), 49–53.

25 "The Man Who Made Pinchot: Dr. B.E. Fernow, 'Father of Forestry,' and His Pioneer Work in This Country and Canada," *New York Times*, 30 January 1910, SM8.

26 "Destructive Agent," *The Globe*, 21 October 1910, 9.

27 *Toronto Daily Star*, 13 July 1911, 9.

28 "The Fire Swept North," 6.
29 "200 to 500 Lives Estimate of Loss in Porcupine District," *Toronto Daily Star*, 13 July 1911, 6.
30 "Vigorous Spirit Show Cochrane Is No Place for Pessimists These Days," *The Globe*, 17 July 1911, 1.
31 Bryan E.C. Bogdanski, "The Rise and Fall of the Canadian Pulp and Paper Sector," *The Forestry Chronicle* 90, no. 6 (November–December 2014): 786.
32 "We Were Warned," *The Globe*, 6 October 1922, 6.
33 "Fire And Fur," *The Globe*, 20 December 1922, 8.
34 S. Lawrence, "Forest and Fires," *The Globe*, 12 October 1922, 4.
35 "How the Big Fires Started," *The Globe*, 20 October 1922, 4.
36 "Declares Fire in Bush Areas People's Fault," *The Globe*, 6 October 1922, 13.
37 "Foresters Discuss Various Methods to Combat Fires, *The Globe*, 10 August 1923, 3.
38 Edith B. Henderson, "Our Heritage, The Forest," *The Globe*, 10 May 1924, 20.
39 "Advises Conference to Plan Protection of Canada's Forests" *The Globe*, 10 September 1924, 20.
40 "Western Provinces Given Full Control of Their Resources," *The Globe*, 2 October 1930, 5.
41 Kalman Seigel, "Forest 'Fire' Loses to 7-State Defense: An Interstate Cooperative Movement to Fight Forest Fires," *New York Times*, 6 October 1951, 6.
42 Harrison E. Salisbury, "The Breath of the Black Dragon in Russia and China," *New York Times*, 1 October 1988, 27.
43 Thomas Walkom, "Climate Change Should Be Linked to Alberta Inferno," *Toronto Star*, 9 May 2016), A11.
44 W.M. Jolly, M.A. Cochrane, P.H. Freeborn, et al., "Climate-Induced Variations in Global Wildfire Danger from 1979 to 2013," *Nature Communications* 6 (2015), 7537.
45 Robin Gill, "'It's Real in My Books,' Lytton B.C. Left to Rebuild after Devastating Fire amid Climate Crisis," *Global News*, 27 August 2021, https://globalnews.ca/news/8136842/its-real-in-my-books-lytton-b-c-left-to-rebuild-after-devastating-fire-amid-climate-crisis/.

Conclusion

1 Ed Struzik, "The Future of Fire in Canada," *The Tyee*, 5 July 2021.
2 Ibid.
3 John T. Fasullo, "Coupled Climate Responses to Recent Australian Wildfire and COVID-19 Emissions Anomalies Estimated in CESM2," *Earth and Space Science, Geophysical Letters*, 15 April 2021.
4 David M. Romps, Jacob T. Seeley, David Vollaro, and John Molinar, "Projected Increase in Lightning Strikes in the United States due to Global Warming," *Science* 3456, no. 6211 (November 2014): 851–4.
5 Daniel Swain, "'Ridiculously Resilient Ridge,' Climate Change and the Future of California's Water," Water Education Foundation, 9 February 2018, https://www.watereducation.org/western-water/ridiculously-resilient-ridge-climate-change-and-future-californias-water.

Index